Progress in Theoretical Computer Science

Leo Bachmair

Canonical Equational Proofs

1991

Birkhäuser
Boston • Basel • Berlin

Leo Bachmair
Department of Computer Science
State University of New York at Stony Brook
Stony Brook, NY 11794
USA

Bachmair, Leo.
    Canonical equational proofs / by Leo Bachmair.
        p.   cm. -- (Progress in theoretical computer science)
    Includes bibliographical references and index.

    1. Rewriting systems (Computer science) 2. Equations. I. Title.
II. Series.
QA267.B32   1991   91-11461
5113.3--dc20   CIP

Printed on acid-free paper.

© Birkhäuser Boston 1991
Softcover reprint of the hardcover 1st edition 1991

ISBN-13:978-0-8176-3555-8   e-ISBN-13:978-1-4684-7118-2
DOI: 10.1007/978-1-4684-7118-2

Camera-ready text prepared in LaTeX by the author.

9 8 7 6 5 4 3 2 1

TO KARIN

Preface

Equations occur in many computer applications, such as symbolic computation, functional programming, abstract data type specifications, program verification, program synthesis, and automated theorem proving. Rewrite systems are directed equations used to compute by replacing subterms in a given formula by equal terms until a simplest form possible, called a normal form, is obtained. The theory of rewriting is concerned with the computation of normal forms. We shall study the use of rewrite techniques for reasoning about equations.

Reasoning about equations may, for instance, involve deciding whether an equation is a logical consequence of a given set of equational axioms. Convergent rewrite systems are those for which the rewriting process defines unique normal forms. They can be thought of as non-deterministic functional programs and provide reasonably efficient decision procedures for the underlying equational theories. The Knuth-Bendix completion method provides a means of testing for convergence and can often be used to construct convergent rewrite systems from non-convergent ones. We develop a proof-theoretic framework for studying completion and related rewrite-based proof procedures.

We shall view theorem provers as proof transformation procedures, so as to express their essential properties as proof normalization theorems. Applying this methodology to equational reasoning methods, we formulate the standard Knuth-Bendix completion procedure as an equational inference system, the inference rules of which represent elementary computation steps that can be combined in different ways to yield a wide range of specific completion procedures. The application of inference rules is reflected by proof transformations on equational proofs and the goal of completion is to deduce enough rewrite rules so that any equational proof can be transformed to a normal-form proof, called a rewrite proof. We will show how proof normalization results can be established via suitable orderings on equational proofs. This approach allows us to abstract from details pertaining to the strategy controlling the inference mechanism. It facilitates comparatively simple and intuitive correctness proofs and enable us to characterize, via a notion of fairness, those strategies that yield correct completion procedures.

We apply this inference system *cum* proof complexity approach to different variants and refinements of completion, including critical pair criteria, extended completion, and ordered completion. Furthermore, we discuss the application of rewrite techniques to inductive theorem proving and describe a method based on the concept of proof by consistency.

This monograph is based on the author's Ph.D. dissertation (submitted to the University of Illinois at Urbana-Champaign in May 1987) and on the following published papers: the inference system *cum* proof complexity approach was first described in Bachmair, Dershowitz and Hsiang (1986); critical pair criteria are discussed in Bachmair and Dershowitz (1988); extended completion is the topic of Bachmair and Dershowitz (1989); an ordered completion procedure is presented in Bachmair, Dershowitz, and Plaisted (1989); and the proof by consistency method is outlined in Bachmair (1988).

Acknowledgements

Most of the research was done while I was at the University of Illinois at Urbana-Champaign and at the State University of New York at Stony Brook. A preliminary version of the manuscript was prepared while I was a visiting professor at the University of Nancy in the summer of 1988. The work was also supported during various stages by the National Science Foundation under grants DCR-8513417 and CCR-8901322.

I wish to take this opportunity to sincerely thank Nachum Dershowitz, my thesis advisor. It has been a great pleasure to work with him.

I also like to thank Jean-Pierre Jouannaud, Jieh Hsiang, and David Plaisted, who read most of the manuscript, for their valuable criticism.

I am grateful to all other colleagues and friends from whom I received comments on various versions of part of this work. They include: Jürgen Avenhaus, Maria-Paola Bonacina, Bruno Buchberger, Jean Gallier, Harald Ganzinger, Miki Hermann, Pierre Lescanne, Uday Reddy, Jean-Luc Rémy, Michael Rusinowitch, G. Sivakumar, Wayne Snyder, the students in my seminars on rewriting, and anonymous reviewers.

Finally, I would like to thank Ron Book, the editor of this series, for his patience.

Contents

1. EQUATIONAL PROOFS

1.1. Introduction

Equations occur in many computer applications, such as symbolic computation, functional programming, abstract data type specifications, program verification, program synthesis, and automated theorem proving. Rewrite systems are collections of directed equations (rewrite rules) used to compute by replacing subterms in a given formula by equal terms until a simplest form possible (a normal form) is obtained. Many formula manipulation systems, such as REDUCE or MACSYMA use equations in this manner. As a computational formalism, rewrite systems have the full power of Turing machines and various equational programming languages have been proposed that are based on the paradigm of rewriting (e.g., O'Donnell 1985; Goguen and Meseguer 1986). We shall study the application of rewrite techniques to reasoning about equations.

The theory of rewriting is a theory of normal forms. Important properties of a rewrite system are termination, which guarantees the existence of normal forms, and the Church-Rosser property, which ensures the uniqueness of normal forms. Rewrite systems that enjoy both properties are called convergent. They can be thought of as non-deterministic functional programs.[1]

For example, the append operation on lists can be defined by the following equations:

$$
\begin{aligned}
app(nil, x) &\approx x \\
app(cons(x, y), z) &\approx cons(x, app(y, z))
\end{aligned}
$$

The two equations define a convergent rewrite system. Any variable-free term can be rewritten to a normal-form term built from the symbols *cons* and *nil*. For example, we have

$$
\begin{aligned}
app(cons(a, nil), cons(b, cons(c, nil))) & \\
\rightarrow_R \quad cons(a, app(nil, cons(b, cons(c, nil)))) & \\
\rightarrow_R \quad cons(a, cons(b, cons(c, nil))) &
\end{aligned}
$$

---

[1] An important aspect of rewriting, which we will not discuss, is the efficient computation of normal forms; see the survey by Klop (1990b).

1

where R denotes the above rewrite system.

Another example of a convergent rewrite system is the following set of equations:

$$
\begin{aligned}
ack(0, x) &\approx S(x) \\
ack(S(x), 0) &\approx ack(x, S(0)) \\
ack(S(x), S(y)) &\approx ack(x, ack(S(x), y))
\end{aligned}
$$

which defines Ackermann's function.

The characterization of free groups by the equations

$$
\begin{aligned}
e \cdot x &\approx x \\
x^- \cdot x &\approx e \\
(x \cdot y) \cdot z &\approx x \cdot (y \cdot z)
\end{aligned}
$$

does not represent a convergent rewrite system, though. For instance, the term $(x^{--} \cdot x^-) \cdot x$ has two different normal forms $x^{--} \cdot e$ and x.

Reasoning about equations may, for instance, involve deciding whether an equation is a logical consequence of a given set of equational axioms. Validity in theories axiomatized by (universally quantified) equations can be characterized proof-theoretically: an equation $s \approx t$ is true in all models of a (countable) set of equations E if and only if it can be proved from E by application of substitutions and replacement of subterms by equal terms. Validity in equational varieties is thus semi-decidable. Convergent rewrite systems provide decision procedures: an equation $s \approx t$ is valid in the equational variety described by a convergent rewrite system if and only if s and t can be rewritten to identical normal forms. Deciding validity in such theories is thus reasonably efficient.

Knuth and Bendix (1970) devised an effective test for deciding whether a given terminating rewrite system is convergent. They also suggested a method with which a convergent system can often be constructed from a non-convergent one.[2] This completion procedure is supplied with a set of equations and a well-founded ordering (a reduction ordering) which is used to orient equations into rewrite rules. If a rewrite system is not convergent, new equations (critical pairs) are deduced by a restricted version of paramodulation, called superposition. More precisely, critical pairs are obtained by paramodulating left-hand sides of rewrite rules into other left-hand sides, with no paramodulations taking place within the variable part of a rule. The given reduction ordering thus provides strong guidance to the deduction mechanism and drastically limits the search space of equational consequences that need to be computed. (The functional reflexivity axioms, which present a problem to many equational theorem provers, need not be included either.)

---

[2] A convergent rewrite system for group theory is given in Section 2.2.

An intriguing feature of completion is that rewrite rules can be used to simplify other, already deduced equations. That is, all terms can be systematically rewritten to normal form by the current rewrite rules. As a consequence, many redundant equations are discarded. While such simplification accounts for the practicality of completion, it also complicates the task of verifying that a procedure is correct (i.e., does in fact produce a convergent rewrite system). We develop a proof-theoretic framework for reasoning about completion and other rewrite-based deduction methods. (Several such methods are outlined in Buchberger 1987.)

Theorem provers can often be viewed as proof transformation or proof normalization procedures, and their essential properties can then be expressed as proof normalization theorems. We shall apply this approach to rewrite-based equational reasoning methods and, in particular, will formulate completion as an equational inference system. The individual inference rules represent elementary computation steps of completion and can be combined in different ways to yield a wide range of completion procedures. This approach allows us to abstract from details pertaining to the strategy controlling the inference mechanism. The application of an inference rule is reflected by certain proof transformations on equational proofs and the goal of completion is to deduce enough rewrite rules so that any equational proof can be transformed to a certain normal-form proof, called a rewrite proof. We will show how proof normalization results can be established via suitable orderings on equational proofs. Proof orderings facilitate comparatively simple and intuitive proofs of correctness and enable us to characterize, via a notion of fairness, those strategies that yield correct completion procedures.

In Chapter 2, we introduce the basic concepts of this inference system *cum* proof complexity approach by applying it to the standard Knuth-Bendix completion procedure. We present an abstract completion inference system and prove the correctness of a wide class of completion procedures. Furthermore, we utilize proof orderings in describing refinements of standard completion based on sorting out redundant critical pairs. These techniques, which are called critical pair criteria, permit better control over the number of equations generated and may considerably improve the efficiency of completion.

Many equational theories cannot be represented as terminating (or convergent) systems, e.g., theories containing commutativity axioms. Instead, problematic axioms such as commutativity can often be built into the completion procedure itself via generalized matching and unification algorithms and a corresponding notion of extended rewriting. Completion procedures based on extended rewriting were first described for the case of associativity and commutativity. In Chapter 3, we describe these and other, more general completion procedures by inference rules. We establish the correctness of various approaches to extended completion: the left-linear rule method, the extended rule method, and the protected rule method. As a special case, we discuss associative-commutative completion in some detail. A brief discussion of extended critical pair criteria is also included.

Completion must be supplied with a reduction ordering which it uses to orient equations into rewrite rules, and may fail if an equation is generated that cannot be oriented. In fact, standard completion may fail even when a convergent system exists and it is supplied with a suitable ordering! In Chapter 4, we address this problem by presenting a method for dealing with unorientable equations, called ordered completion, which is designed to construct sets of equations that define unique normal forms for variable-free terms only. This weaker normal-form property is sufficient in many applications, such as refutational theorem proving. Ordered completion is a refutationally complete theorem prover for equational theories, but has the advantage over paramodulation that terms can always be kept in fully-simplified form and that fewer equational consequences need to be considered, as the ordering supplied to the procedure gives some measure of direction to the prover. Ordered completion is guaranteed to find a convergent system, if one exists and the reduction ordering supplied to the procedure satisfies some reasonable conditions. In the more general context of Horn clauses with equality it yields an inference system consisting of restricted versions of positive unit resolution and paramodulation (with simplification).

In many applications, such as algebraic data type specifications and equational programming, equations are intended to define a certain standard model, called the initial model. This initial algebra semantics typically requires proof methods that employ some induction schema, e.g., induction on the structure of terms. In Chapter 5, we discuss an alternative approach based on the concept of proof by consistency and apply it to equational theories that are presented as rewrite systems defining unique normal forms for variable-free terms. In contrast with so-called inductive completion procedures, this proof by consistency method can handle unorientable equations. The method is linear in that every deduction step involves one of the original axioms. It is refutationally complete in that it refutes any equation which is not an inductive theorem.

We assume that the reader is familiar with the fundamental concepts of term rewriting; a brief introduction is provided in the remaining sections of this chapter. Our terminology is consistent with Dershowitz and Jouannaud (1990). For further details on rewriting the reader may also consult the surveys by Huet and Oppen (1980), Klop (1987, 1990a), and Avenhaus and Madlener (1990).

1.2. Terms

Let $\mathcal{F}$ and $\mathcal{V}$ be two disjoint (countable) sets. The elements of $\mathcal{F}$ are called *function symbols*; the elements of $\mathcal{V}$, *variables*. With each function symbol f we associate a non-negative number, called its *arity*. Function symbols of arity 0 are called *constants*.

A *term* is either a variable or an expression $ft_1 \ldots t_n$, where f is a function symbol of arity n and $t_1, \ldots, t_n$ are terms. The set of all terms built from function symbols in $\mathcal{F}$ and variables in $\mathcal{V}$ is denoted by $T(\mathcal{F}, \mathcal{V})$. Terms containing no variables are called *ground terms*. The set of all ground terms is denoted by $T(\mathcal{F})$. To enhance readability, we shall often use parentheses and infix notation. For example, we may write $(S(x) + y) + 0$ instead of $++Sxy0$. The term $(S(x) + y) + 0$ is non-ground, whereas $S(0) + 0$ is ground.

A term s is said to be a *subterm* of a term t if either $s = t$, or else $t = ft_1 \ldots t_n$ and s is a subterm of one of the terms t_i. By a *proper subterm* of t we mean a subterm distinct from t.

We use sequences of positive numbers, also called *positions*, to refer to specific subterms in a term. The empty sequence λ is a position in any term t, whereas a sequence ip is a position in a term $t = ft_1 \ldots t_n$ only if $1 \leq i \leq n$ and p is a position in t_i. (The *concatenation* of two sequences p and q is denoted by juxtaposition pq.) If p is a position in a term t then the *subterm* $t|_p$ of t at position p is t, if $p = \lambda$, and $t_i|_q$, if $t = ft_1 \ldots t_n$ and $p = iq$, for some i with $1 \leq i \leq n$. We say that a position p is *below* a position q (or q is *above* p) if $p = qq'$, for some position q'. If q' is non-empty, then p is said to be *strictly below* q (and q *strictly above* p). Two positions are *disjoint* if neither one is below the other.

The result of *replacing* in t the subterm at position p by s is denoted by $t[s]_p$ and is defined to be the term s, if $p = \lambda$, and the term $ft_1 \ldots t_{i-1}t_i[s]_q t_{i+1} \ldots t_n$, if $t = ft_1 \ldots t_n$ and $p = iq$. More generally, if $p_1, \ldots, p_n$ are pairwise disjoint positions, for some $n \geq 2$, we write $t[s_1, \ldots, s_n]_{p_1, \ldots, p_n}$ to denote the term $(t[s_1]_{p_1})[s_2, \ldots, s_n]_{p_2, \ldots, p_n}$. We also write $t[s]$ to indicate that the term t contains s as a subterm, and (ambiguously) denote by $t[u]$ the result of replacing a particular occurrence of s in t by u.

A *substitution* is a mapping from variables to terms. The value of a substitution σ for variable x is denoted by $x\sigma$. A substitution σ can be uniquely extended to a mapping on terms in such a way that $(ft_1 \ldots t_n)\sigma = f(t_1\sigma) \ldots (t_n\sigma)$, for all terms $ft_1 \ldots t_n$. The composition of two substitutions σ and τ is denoted by juxtaposition $\sigma\tau$. That is, $t\sigma\tau = (t\sigma)\tau$ for all terms t. We usually need to consider only substitutions for which $x\sigma = x$ for all but a finite number of variables. By $\{x_1 \mapsto t_1, \ldots, x_n \mapsto t_n\}$ we denote the substitution σ for which $x\sigma = t_i$, if $x = x_i$, for some i with $1 \leq i \leq n$, and $x\sigma = x$, otherwise.

A term t is an *instance* of (or *matches*) another term s if $t = s\sigma$, for some substitution σ. An instance t of s is said to be *proper* if s is not an instance of t. Thus $-x + 0$ and $x + x$ are proper instances of $x + y$, while $x + z$ is a non-proper instance. Two terms s and t are *literally similar* if they are instances of each other. In other words, two terms are literally similar if and only if they can be obtained from each other by a suitable renaming of variables. The two terms $x + y$ and $x + z$ are literally similar, but $x + y$ and $x + x$ are not.

Two terms s and t are said to be *unifiable* if there exists a substitution σ, called a *unifier*, such that $s\sigma = t\sigma$. For example, the two terms fx and fgy are unifiable, whereas fx and gy are not. The two substitutions $\{x \mapsto gy\}$ and $\{x \mapsto gfz, y \mapsto fz\}$ are both unifiers of fx and fgy, though the first unifier is more general than the second. More precisely, a unifier σ of s and t is said to be *most general* if for every unifier τ (of s and t) there exists a substitution τ' such that $x\sigma\tau' = x\tau$, for all variables x. Unifiability of terms is decidable. Furthermore, if two terms are unifiable, then they have a most general unifier, which is unique up to renaming of variables. Algorithms for computing the most general unifier of two terms (if one exists) have been described by Robinson (1965), Paterson and Wegman (1978), and Martelli and Montanari (1982). For a recent survey on unification see Jouannaud and Kirchner (1990).

1.3. Equations

An *equation* is a pair of terms, written $s \approx t$. We are interested in various congruence and rewrite relations induced by a given set of equations. An equivalence relation[3] $\sim$ on terms is called a *congruence* if $s_1 \sim t_1, \ldots,$ $s_n \sim t_n$ implies $fs_1 \ldots s_n \sim ft_1 \ldots t_n$, for all terms $s_1, \ldots, s_n, t_1, \ldots, t_n$ and function symbols f of arity n. A binary relation $\rightarrow$ on terms is called a *rewrite relation* if $s \rightarrow t$ implies $u[s\sigma]_p \rightarrow u[t\sigma]_p$, for all terms s, t, and u, positions p in u, and substitutions σ. Rewrite relations are often denoted by arrows of one kind or another. If $\rightarrow$ is a binary relation, we denote by $\leftarrow$ its inverse; by $\leftrightarrow$ its symmetric closure $\rightarrow \cup \leftarrow$; by $\rightarrow^+$ its transitive closure; by $\rightarrow^*$ its transitive-reflexive closure; and by $\leftrightarrow^*$ its symmetric-transitive-reflexive closure.

If E is a set of equations, we write $s \rightarrow_E t$ to indicate that there exist a term w, a position p in w, a substitution σ, and an equation $u \approx v$ in E, such that $s = w[u\sigma]_p$ and $t = w[v\sigma]_p$. The relation $\rightarrow_E$ is called the

---

[3] An *equivalence relation* is a reflexive, transitive, and symmetric binary relation.

rewrite relation induced by E. We also say that s *rewrites* to t (in one step) if $s \rightarrow_E t$. A term that cannot be rewritten is said to be in *normal form* or *irreducible* (by E). A *normal form* of t (in E) is any irreducible term u for which $t \rightarrow_E^* u$. We write $s \leftrightarrow_E t$ if either $s \rightarrow_E t$ or $t \rightarrow_E s$.

The symmetric-transitive-reflexive closure $\leftrightarrow_E^*$ of $\rightarrow_E$ is called the *equational theory* induced by E. We also say that two terms s and t are *equivalent* in E if $s \leftrightarrow_E^* t$. A set E of equations will be called a *rewrite system* if the corresponding rewrite relation $\rightarrow_E$ is the primary object of study. The equations of a rewrite system are also called *rewrite rules*.

A binary relation $\rightarrow$ is said to be *Church-Rosser* if for any two elements s and t with $s \leftrightarrow^* t$, there exists an element v such that $s \rightarrow^* v$ and $v \leftarrow^* t$. It is said to be *terminating* if there is no infinite sequence $t_0 \rightarrow t_1 \rightarrow t_2 \cdots$. Terminating Church-Rosser relations are called *convergent*.

A rewrite system R is called convergent if the corresponding rewrite relation $\rightarrow_R$ is convergent. Convergent rewrite relations define unique normal forms. Thus, if R is a finite convergent rewrite system, then equivalence of terms in R is decidable: two terms are equivalent in R if and only if they can be rewritten to identical normal forms.

Finally, a rewrite system R is said to be *reduced* if for all rewrite rules $s \approx t$ in R, the term s is irreducible by $R \setminus \{s \approx t\}$ and t is irreducible by R. Reduced convergent rewrite systems are often called *canonical*. They are unique up to renaming of variables (Metivier 1983).

1.4. Orderings

An (strict partial) *ordering* is an irreflexive, transitive binary relation; a *quasi-ordering*, a reflexive, transitive binary relation. For example, the greater-than relation on the natural numbers is a strict ordering, while the greater-than-or-equal-to relation is a quasi-ordering.

If $\succ$ is a strict ordering, then its reflexive closure $\succeq$ is a quasi-ordering. On the other hand, if $\succsim$ is a quasi-ordering, the corresponding equivalence $\sim$ and strict ordering $\succ$ are defined as follows: $s \sim t$ if $s \succsim t$ and $t \succsim s$; and $s \succ t$ if $s \succsim t$ and $t \not\succsim s$. An ordering $\succ$ is called *well-founded* if it is terminating. A quasi-ordering $\succsim$ is said to be well-founded if its strict part $\succ$ is well-founded.

If $\succ_1$ and $\succ_2$ are both orderings, than their *lexicographic combination* is an ordering on pairs, defined by: $(s, t) \succ (s', t')$ if either $s \succ_1 s'$ or else $s = s'$ and $t \succ_2 t'$. More generally, if $\succsim_1$ is a quasi-ordering, we define: $(s, t) \succ (s', t')$ if either $s \succ_1 s'$ or else $s \sim_1 s'$ and $t \succ_2 t'$, where $\succ_1$ and $\sim_1$ are the strict ordering and equivalence, respectively, associated with $\succsim_1$. The lexicographic combination of more than two orderings (or quasi-orderings) is defined in the obvious way. A lexicographic ordering is well-founded if and only if all its component orderings are well-founded.

A *multiset* over a set S is a mapping M from S to the natural numbers. Intuitively, $M(x)$ specifies the number of occurrences of x in M. We say that x is an *element* of M if $M(x) > 0$. The *union* and *intersection* of multisets are defined in the usual way by the identities $M_1 \cup M_2(x) = M_1(x) + M_2(x)$ and $M_1 \cap M_2(x) = \min(M_1(x), M_2(x))$. A multiset M is *finite* if the set $\{x : M(x) > 0\}$ is finite. For simplicity, we often use a set-like notation to describe (finite) multisets. For example, $\{x, x, x\}$ denotes the multiset M for which $M(x) = 3$ and $M(y) = 0$, for $y \neq x$.

Any partial ordering $\succ$ on a set S can be extended to an ordering $\succ_{mul}$ on (finite) multisets over S as follows: $M \succ_{mul} N$ if (i) $M \neq N$ and (ii) whenever $N(x) > M(x)$ then $M(y) > N(y)$, for some y such that $y \succ x$. In other words, according to the multiset ordering any element of a multiset can be replaced by any finite number of smaller elements. Dershowitz and Manna (1979) showed that the multiset ordering $\succ_{mul}$ is well-founded (on finite multisets) if and only if the ordering $\succ$ is well-founded.

A *rewrite ordering* is an ordering that is also a rewrite relation. Well-founded rewrite orderings are called *reduction orderings*. We say that a reduction ordering $\succ$ satisfies the *subterm property* if $t[s] \succ s$, for all terms t and proper subterms s of t. Reduction orderings that satisfy the subterm property are called *simplification orderings*.

A rewrite relation $\rightarrow_R$ terminates if and only if the rewrite system R is contained (as a subset) in some reduction ordering. If a convergent rewrite system R is contained in a reduction ordering $\succ$, then a term is irreducible with respect to R if and only if it is minimal (with respect to $\succ$) in its equivalence class. The following orderings, which are based on lexicographic or multiset orderings, are particularly useful for proving the termination of rewrite relations.

Let f be a function symbol of arity n with which we associate a mapping (called the *status* of f) that assigns to each ordering $\succ$ on terms an ordering $\succ^f$ on n-tuples of terms. In particular, the function symbol f is said to have *multiset status*, if $\succ^f$ is defined by: $(s_1, \ldots, s_n) \succ^f (t_1, \ldots, t_n)$ if $\{s_1, \ldots, s_n\} \succ_{mul} \{t_1, \ldots, t_n\}$; while it is said to have *lexicographic status* if there exists a permutation π of $\{1, \ldots, n\}$, such that $(s_1, \ldots, s_n) \succ^f (t_1, \ldots, t_n)$ if (i) $(s_{\pi(1)}, \ldots, s_{\pi(n)}) \succ_{lex} (t_{\pi(1)}, \ldots, t_{\pi(n)})$ and (ii) $f s_1 \ldots s_n \succ t_i$, for all i with $1 \leq i \leq n$. (Here $\succ_{lex}$ denotes the n-fold lexicographic combination of the ordering $\succ$.)

Let $\succ$ be an ordering, called a *precedence*, on a given set of function symbols, and suppose that each function symbol has either multiset or lexicographic status. Then the corresponding *recursive path ordering* $\succ_{rpo}$

is recursively defined by:

$$s = f s_1 \ldots s_m \succ_{rpo} g t_1 \ldots t_n = t$$

if

$$s_i \succeq_{rpo} t, \quad \text{for some } i \text{ with } 1 \le i \le m,$$

or

$$f \succ g \text{ and } s \succ_{rpo} t_j, \quad \text{for all } j \text{ with } 1 \le j \le n,$$

or

$$f = g \text{ and } (s_1, \ldots, s_m) \succ_{rpo}^f (t_1, \ldots, t_m)$$

(Kamin and Lévy 1980).

A recursive path ordering is more specifically called a *multiset path ordering* and denoted by $\succ_{mpo}$, if each function symbol has multiset status (Dershowitz 1982a). A *lexicographic path ordering* $\succ_{lpo}$ is a recursive path ordering where each function symbol has lexicographic status with π being the identity permutation.

For example, if $* \succ +$ in the given precedence, then $(a + b) * c \succ_{mpo} (a * c) + (b * c)$, as both $(a + b) * c \succ_{mpo} (a * c)$ and $(a + b) * c \succ_{mpo} (b * c)$. The two terms $(a + b) + c$ and $a + (b + c)$ are not comparable in the multiset path ordering, while $(a + b) + c \succ_{lpo} a + (b + c)$.

Proposition 1.1. (Kamin and Lévy 1980) *Any recursive path ordering is a simplification ordering if the underlying precedence on function symbols is well-founded.*

For a detailed discussion of termination orderings the reader may consult the survey by Dershowitz (1987).

1.5. Proofs

Theorem provers can often be viewed as proof transformation or normalization procedures. Such a theorem prover is said to be correct if enough consequences can be deduced so that any arbitrary proof can be transformed to a normal-form proof. We study rewrite-based equational reasoning methods from this point of view. In this context, the set of theorems corresponds to some congruence relation induced by a given set of equations; proofs are finite sequences of equational replacements; and normal-form proofs are those proofs in which equations are used in a specified direction, as one-way rewrite rules.

In describing equational proof procedures, we shall assume that equations are labelled by non-negative numbers or the symbols $\perp$ or $\top$. We write $s \approx_n t$ to denote labelled equations.[4]

Definition 1.2. A *(equational) proof step* is an expression $s \leftrightarrow^p_e t$, where s and t are terms, e is an equation $u \approx_n v$, and p is a position in s, such that $s|_p = u\sigma$ and $t = s[v\sigma]_p$, for some substitution σ. We say that $s \leftrightarrow^p_{u \approx_n v} t$ is a proof step in E if either $u \approx_n v$ or $v \approx_n u$ is an equation in E.

Occasionally we will write $s \simeq_n t$ to ambiguously denote $s \approx_n t$ or $t \approx_n s$. Evidently, there is a proof step $s \leftrightarrow^p_{u \approx_n v} t$ in E if and only if $s \leftrightarrow_E t$.

Definition 1.3. A *(equational) proof* (of $t_0 \approx t_n$) is any finite sequence of proof steps

$$t_0 \leftrightarrow^{p_1}_{e_1} t_1, t_1 \leftrightarrow^{p_2}_{e_2} t_2, \ldots, t_{n-1} \leftrightarrow^{p_n}_{e_n} t_n,$$

usually written in abbreviated form,

$$t_0 \leftrightarrow^{p_1}_{e_1} t_1 \leftrightarrow^{p_2}_{e_2} t_2 \leftrightarrow^{p_3}_{e_3} \cdots t_{n-1} \leftrightarrow^{p_n}_{e_n} t_n.$$

The empty sequence serves as a proof of any "trivial" equation $t \approx t$.

The letters P and Q are used to denote proofs. We say that P is a proof in E if each proof step of P is in E. An equation $s \approx t$ is provable in E, in this sense, if and only if $s \leftrightarrow^*_E t$. For simplicity, we often denote by $s \rightarrow_E t$ any proof step $s \leftrightarrow^p_{u \approx_n v} t$, where $u \approx_n v$ is an equation in E; by $s \leftarrow_E t$, a proof step $s \leftrightarrow^p_{u \approx_n v} t$, where $v \approx_n u$ is an equation in E; and by $s \leftrightarrow_E t$ a proof step $s \rightarrow_E t$ or $s \leftarrow_E t$. We also write $s \leftrightarrow^*_E t$ to denote arbitrary proofs in E. A proof of the form $s \leftarrow_E u \rightarrow_E t$ is called a *peak*; a proof $t_0 \rightarrow^*_E t_k \leftarrow^*_E t_n$, a *rewrite proof*.

If e_i is an equation $u_i \approx_{k_i} v_i$, for $1 \leq i \leq n$, and P is a proof

$$t_0 \leftrightarrow^{p_1}_{e_1} t_1 \leftrightarrow^{p_2}_{e_2} t_2 \cdots t_{n-1} \leftrightarrow^{p_n}_{e_n} t_n,$$

we denote by P^{-1} the proof

$$t_n \leftrightarrow^{p_n}_{v_n \approx_{k_n} u_n} t_{n-1} \cdots t_2 \leftrightarrow^{p_2}_{v_2 \approx_{k_2} u_2} t_1 \leftrightarrow^{p_1}_{v_1 \approx_{k_1} u_1} t_0;$$

by $P\sigma$ the proof

$$t_0\sigma \leftrightarrow^{p_1}_{e_1} t_1\sigma \leftrightarrow^{p_2}_{e_2} t_2\sigma \cdots t_{n-1}\sigma \leftrightarrow^{p_n}_{e_n} t_n\sigma;$$

---

[4] The labels allow us to make finer distinctions between equational proofs. For instance, we will employ labels to indicate whether an equation is to be used in a specified direction as a (one-way) rewrite rule. All notions defined for unlabelled equations will also be applied to labelled equations. For instance, when speaking of the rewrite relation induced by a set of labelled equations E, we mean the rewrite relation induced by the set of equations $\{s \approx t : s \approx_n t \in E,$ for some $n\}$.

and by $u[P]_q$, where q is a position in u, the proof

$$u[t_0]_q \leftrightarrow^{qp_1}_{e_1} u[t_1]_q \leftrightarrow^{qp_2}_{e_2} u[t_2]_q \cdots u[t_{n-1}]_q \leftrightarrow^{qp_n}_{e_n} u[t_n]_q.$$

If P is a proof (in E), then P^{-1}, $P\sigma$, and $u[P]_q$ are also proofs (in E). We speak of a *ground proof* if all terms $t_0, \ldots, t_n$ are ground. By a *subproof* of P we mean any proof

$$t_i \leftrightarrow^{p_{i+1}}_{e_{i+1}} t_{i+1} \cdots t_{j-1} \leftrightarrow^{p_j}_{e_j} t_j,$$

where $0 \le i \le j \le n$. We write $P[Q]$ to indicate that P contains Q as a subproof, and denote by $P[Q']$ the proof obtained from P by replacing Q by Q'. The *composition* of a proof P (of $s \approx t$) and a proof Q (of $t \approx u$) is denoted by juxtaposition PQ (and is a proof of $s \approx u$).

Definition 1.4. A *proof rewrite relation* is any binary relation $\Rightarrow$ on proofs such that (i) $P \Rightarrow Q$ implies $u[P\sigma]_q \Rightarrow u[Q\sigma]_q$, for all proofs P and Q, substitutions σ, terms u, and positions q in u; and (ii) $Q \Rightarrow Q'$ implies $P[Q] \Rightarrow P[Q']$, for all proofs $P[Q]$, Q, and Q'. If in addition (iii) P and Q are proofs of the same equation whenever $P \Rightarrow Q$, then the proof rewrite relation is called a *(proof) transformation relation*.

If a proof rewrite relation is a well-founded ordering, it is called a *proof reduction ordering* (or simply *proof ordering*).

2. STANDARD COMPLETION

Knuth and Bendix (1970) proposed a procedure that attempts to transform a given set of equations into a convergent rewrite system. This completion procedure must be supplied with a reduction ordering, which determines the direction in which an equation is to be *oriented* into a rewrite rule. It *deduces* new equations by a process called superposition. An intriguing feature of the procedure is that rewrite rules can be used to *simplify* other, already deduced equations. Terms can therefore be kept in fully simplified form and redundant equations can be discarded. Deduction and simplification are the two main components of completion. While simplification accounts for the practicality of completion, it also complicates the task of verifying that a procedure is correct (i.e., does in fact produce a convergent set of equations).

We reformulate the Knuth-Bendix completion method as an equational inference system and demonstrate that completion can be viewed as a process of proof simplification, the goal of which is to deduce enough rewrite rules so that any equational proof can be transformed to a normal-form proof, that is, a rewrite proof.

In this context, an inference rule is a binary relation on (finite) sets of equations. The inference rules represent elementary computation steps of completion. They can be combined in different ways to yield a wide range of completion procedures. Each inference rule induces certain proof transformations that can be described by rewrite rules on equational proofs. We present well-founded orderings on proofs that can be used to establish proof normalization results for (abstract) completion and related rewrite-based proof methods.

In this chapter, equations are assumed to be labelled by one of the symbols $\top$ or $\bot$. We usually leave the labels implicit. Equations $s \approx_\bot t$ are called *rewrite rules* and written $s \to t$. We write $E \, ; R$ to denote a set $E \cup R$, where E is a set of equations $s \approx_\top t$ and R is a set of rules $s \approx_\bot t$. A proof step $s \leftrightarrow_E t$ is called an *equality step*; a proof step $s \to_R t$ or $s \leftarrow_R t$, a *rewrite step*.

13

2.1. Basic Completion

Let $\succ$ be a reduction ordering on terms. The inference system $\mathcal{B}^{\succ}$ consists of the following inference rules:

DEDUCTION:
$$\frac{E\,;R}{E\cup\{s\approx t\}\,;R} \qquad \text{if } s \leftarrow_R u \rightarrow_R t$$

ORIENTATION:
$$\frac{E\cup\{s\simeq t\}\,;R}{E\,;R\cup\{s\rightarrow t\}} \qquad \text{if } s\succ t$$

DELETION:
$$\frac{E\cup\{s\approx s\}\,;R}{E\,;R}$$

SIMPLIFICATION:
$$\frac{E\cup\{s\simeq t\}\,;R}{E\cup\{u\simeq t\}\,;R} \qquad \text{if } s\rightarrow_R u$$

where E and R denote finite sets of equations, with R being contained in the reduction ordering $\succ$.

Inference systems $\mathcal{B}^{\succ}$ are called *basic completion systems*. We usually leave $\succ$ implicit and write $\mathcal{B}$ instead of $\mathcal{B}^{\succ}$.

In essence, new equations $s\approx t$ are obtained by rewriting some term u in two different ways to s and t, respectively. Such equations are deduced from R, but not from E. They serve to eliminate peaks $s\leftarrow_R u\rightarrow_R t$. As we shall see, only equations obtained from so-called critical peaks (of which there are only finitely many, whenever R is finite) need to be deduced.

For example, the two rules $e\cdot x\rightarrow x$ and $(x\cdot y)\cdot z\rightarrow x\cdot(y\cdot z)$ determine a peak $y\cdot z\leftarrow_R (e\cdot y)\cdot z\rightarrow_R e\cdot(y\cdot z)$, so that the equation $y\cdot z\approx(e\cdot y)\cdot z$ can be deduced. Since $e\cdot(y\cdot z)\rightarrow_R y\cdot z$, the equation can be simplified to a trivial equation $y\cdot z\approx y\cdot z$, which can be deleted.

Orientation of equations into rules is controlled by the given reduction ordering $\succ$ to ensure that only terminating rewrite systems can be derived from any initial rewrite system contained in $\succ$. For instance, the equation $e^-\approx e$ can be oriented (with respect to any simplification ordering) into a rule $e^-\rightarrow e$.

Given an equational inference system $\mathcal{I}$, we write $E\vdash_{\mathcal{I}} E'$ to indicate that E' can be obtained from E by application of an inference rule of $\mathcal{I}$.

Definition 2.1. A (possibly infinite) sequence $E_0\vdash_{\mathcal{I}} E_1\vdash_{\mathcal{I}}\cdots$ is called a *derivation* in $\mathcal{I}$ from E_0. The (*lower*) *limit* $\bigcup_i\bigcap_{j\geq i} E_j$ of a sequence of equations $E_0, E_1,\ldots$ is denoted by E_∞. Equations in E_∞ are called *persisting*.

An inference system $\mathcal{I}$ is said to be *sound* if, for all sets of equations E and E', the same equations are provable in E and E' whenever $E \vdash_\mathcal{I} E'$. Basic completion is sound in this sense.

Lemma 2.2. (Soundness) *If $E\,;R \vdash_B E'\,;R'$, then the congruence relations $\leftrightarrow^*_{E \cup R}$ and $\leftrightarrow^*_{E' \cup R'}$ are the same.*

Proof. Suppose $E\,;R \vdash_B E'\,;R'$ is by application of an inference rule ρ. Evidently, the assertion is true if ρ is a deduction, orientation, or deletion inference. If ρ is a simplification inference, then $E = E'' \cup \{s \simeq t\}$, $E' = E'' \cup \{u \simeq t\}$, and $R' = R$, where $s \to_R u$. We have $s \to_{R'} u \leftrightarrow_{E'} t$ and $u \leftarrow_R s \leftrightarrow_E t$, which implies that the two congruence relations $\leftrightarrow^*_{E \cup R}$ and $\leftrightarrow^*_{E' \cup R'}$ are the same.   Q.E.D.

If $E\,;R \vdash_{B^\succ} E'\,;R'$ and R is contained in the reduction ordering $\succ$, then R' is also contained in $\succ$. Consequently, the rewrite system R_∞ is terminating for any derivation in $B^\succ$ for which the initial rewrite system R_0 is contained in the reduction ordering $\succ$. Furthermore, the inference rules of basic completion never decrease the strength of rewriting. That is, if $E\,;R \vdash_B E'\,;R'$ then any term t that is reducible by R is also reducible by R'.

Definition 2.3. By a *basic completion procedure* we mean a program that accepts as input a reduction ordering $\succ$ and a set of equations $E_0\,;R_0$ where R_0 is contained in $\succ$; and uses the inference rules of $B^\succ$ to generate a derivation from $E_0 \cup R_0$. We say that a completion procedure *succeeds* for a given input if $E_\infty = \emptyset$ and R_∞ is convergent. A procedure *fails* if $E_\infty \neq \emptyset$. It is called *correct* if R_∞ is convergent whenever $E_\infty = \emptyset$.

We shall also distinguish between failing and non-failing derivations depending on whether E_∞ is empty or non-empty. Correctness requires that all non-failing derivations result in convergent rewrite systems. We will present techniques for establishing the correctness, in this sense, of completion procedures.

If an equational theory can be represented as a convergent rewrite system, then equivalence of terms is decidable. Not all equational theories are decidable in this sense, though. An example of an undecidable equational theory is *combinatory logic*:

$$
\begin{array}{rcl}
I \cdot x & \approx & x \\
(K \cdot x) \cdot y & \approx & x \\
((S \cdot x) \cdot y) \cdot z & \approx & (x \cdot z) \cdot (y \cdot z)
\end{array}
$$

In general, a completion procedure may either (i) succeed in constructing a finite convergent system, (ii) fail, or (iii) not terminate and instead compute successive approximations of an infinite convergent system R_∞.

2.2. Proof Transformation

We will use binary relations on (equational) proofs to describe the effect of inferences on the proof level.

Definition 2.4. A *(proof) transformation rule* is a pair of proofs, written $P \Rightarrow Q$, where P and Q are proofs of the same equation. A *(proof) transformation system* is a set of such transformation rules.

For the purpose of transforming proofs we do not distinguish between a proof P and its inverse P^{-1}.

Definition 2.5. Let $\mathcal{R}$ be a proof transformation system. By $\Rightarrow_{\mathcal{R}}$ we denote the smallest (proof) transformation relation that contains $\mathcal{R}$ and for which $P^{-1} \Rightarrow_{\mathcal{R}} Q^{-1}$, whenever $P \Rightarrow_{\mathcal{R}} Q$. A *proof transformation step* by $\mathcal{R}$ is a pair of proofs for which $P \Rightarrow_{\mathcal{R}} Q$. We also say that P can be *transformed* to Q (in zero or more steps) if $P \Rightarrow_{\mathcal{R}}^{*} Q$.

A transformation system $\mathcal{R}$ is called *terminating* if the corresponding transformation relation $\Rightarrow_{\mathcal{R}}$ is terminating. A proof P is in *normal form* (with respect to $\mathcal{R}$) if there is no proof Q such that $P \Rightarrow_{\mathcal{R}} Q$.

Definition 2.6. A proof transformation system $\mathcal{R}$ is said to *reflect* an inference system $\mathcal{I}$ if $E \vdash_{\mathcal{I}} E'$ implies that every proof P in E can be transformed by $\mathcal{R}$ to some proof P' in E'.

Every basic completion system $\mathcal{B}$ is reflected in this sense by a transformation system $\mathcal{R}_{\mathcal{B}}$. For instance, deduction is reflected by transformation rules of the form

$$s \overset{p}{\underset{w \approx_{\perp} v}{\longleftarrow}} u \overset{q}{\underset{v' \approx_{\perp} w'}{\longrightarrow}} t \;\;\Rightarrow\;\; s \overset{\lambda}{\underset{s \approx_{\top} t}{\longleftrightarrow}} t$$

where $v \succ w$ and $v' \succ w'$. Orientation, deletion, and simplification are reflected by transformation rules

$$
\begin{array}{lll}
s \overset{\lambda}{\underset{s \approx_{\top} t}{\longleftrightarrow}} t & \Rightarrow & s \overset{\lambda}{\underset{s \approx_{\perp} t}{\longrightarrow}} t & \text{where } s \succ t \\[4pt]
s \overset{\lambda}{\underset{s \approx_{\top} t}{\longleftrightarrow}} t & \Rightarrow & s \overset{p}{\underset{s' \approx_{\perp} u'}{\longrightarrow}} u \overset{\lambda}{\underset{u \approx_{\top} t}{\longleftrightarrow}} t & \text{where } s' \succ u' \\[4pt]
s \overset{\lambda}{\underset{s \approx_{\top} s}{\longleftrightarrow}} s & \Rightarrow & \square
\end{array}
$$

where $\square$ denotes the empty proof (i.e., the empty sequence of proof steps), cf. Figure 2.1.

The proof transformation system $\mathcal{R}_{\mathcal{B}}^{\succ}$ (or simply $\mathcal{R}_{\mathcal{B}}$ if $\succ$ is clear from the context) is the set of all such proof transformation rules. The transformation relation induced by $\mathcal{R}_{\mathcal{B}}$ is denoted by $\Rightarrow_{\mathcal{B}}$.

Lemma 2.7. (Reflection) *The proof transformation system $\mathcal{R}_{\mathcal{B}}$ reflects basic completion $\mathcal{B}$.*

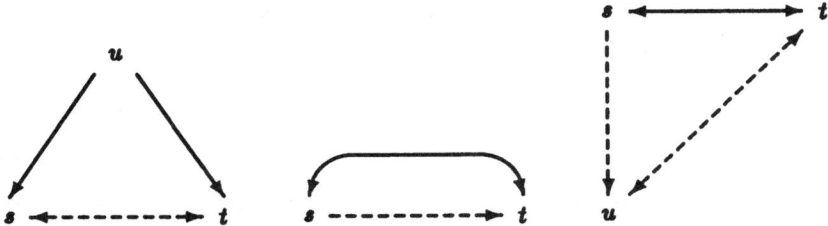

Figure 2.1: *Proof transformation rules*

Proof. If $E\,;R \vdash_{\mathcal B} E'\,;R'$ by deduction, then $E' = E \cup \{s \approx_T t\}$, $R' = R$, and there exists a peak Q of the form $s \leftarrow_R u \rightarrow_R t$. If P is a proof in $E \cup R$, let P' be the proof in $E' \cup R'$ obtained from P by replacing each subproof $w[Q\sigma]_q$ by the proof $w[s\sigma \leftrightarrow^\lambda_{s \approx_T t} t\sigma]_q$. If P contains no subproof subproof $w[Q\sigma]_q$, then $P = P'$. Since each such replacement represents a proof transformation step by $\mathcal R_{\mathcal B}$, we may conclude that $P \Rightarrow^*_{\mathcal B} P'$.

Similar arguments apply to the other inference rules. Q.E.D.

Let us illustrate the proof transformation process with the axioms of *group theory*:

$$
\begin{aligned}
e \cdot x &\approx x \\
x^- \cdot x &\approx e \\
(x \cdot y) \cdot z &\approx x \cdot (y \cdot z).
\end{aligned}
$$

The equation $x^{--} \cdot e \approx x$ is provable:

$$
\begin{aligned}
x^{--} \cdot e \;\;&\leftrightarrow_E\;\; x^{--} \cdot (x^- \cdot x) \\
&\leftrightarrow_E\;\; (x^{--} \cdot x^-) \cdot x \\
&\leftrightarrow_E\;\; e \cdot x \\
&\leftrightarrow_E\;\; x.
\end{aligned}
$$

Let now $\succ_{lpo}$ be the lexicographic path ordering corresponding to a precedence $\succ$ in which $\cdot \succ {}^- \succ e$. By repeated orientation of equations we obtain the following rewrite system R_1:

$$
\begin{aligned}
e \cdot x &\rightarrow x \\
x^- \cdot x &\rightarrow e \\
(x \cdot y) \cdot z &\rightarrow x \cdot (y \cdot z).
\end{aligned}
$$

These inference steps are reflected by a sequence of proof transformation steps $P_0 \Rightarrow_{\mathcal{B}}^+ P_1$, where P_0 is the above proof and P_1 is

$$x^{--} \cdot e \leftarrow_{R_1} \quad x^{--} \cdot (x^- \cdot x) \leftarrow_{R_1} \quad \overset{(x^{--} \cdot x^-) \cdot x}{\longrightarrow_{R_1} e \cdot x} \quad \rightarrow_{R_1} x.$$

The middle two steps of this proof form a peak which is an instance of a more general peak

$$x^- \cdot (x \cdot y) \leftarrow_{R_1} (x^- \cdot x) \cdot y \rightarrow_{R_1} e \cdot y$$

involving the third and second rule of R_1. We can deduce the equation $x^- \cdot (x \cdot y) \approx e \cdot y$ and obtain a new proof P_2:

$$x^{--} \cdot e \leftarrow_{R_2} \quad x^{--} \cdot (x^- \cdot x) \leftrightarrow_{E_2} e \cdot x \quad \rightarrow_{R_2} x,$$

where E_2 consists of the new equation and $R_2 = R_1$. Again we have $P_1 \Rightarrow_{\mathcal{B}}^+ P_2$. The equation $x^- \cdot (x \cdot y) \approx e \cdot y$ can be simplified to $x^- \cdot (x \cdot y) \approx y$ and then oriented. This is reflected by a proof transformation $P_2 \Rightarrow_{\mathcal{B}}^+ P_3$, where P_3 is

$$x^{--} \cdot e \leftarrow_{R_3} \quad \overset{x^{--} \cdot (x^- \cdot x)}{\longrightarrow_{R_3} x \leftarrow_{R_3}} \quad \overset{e \cdot x}{\longrightarrow_{R_3} x.}$$

The "trivial" peak $x \leftarrow_{R_3} e \cdot x \rightarrow_{R_3} x$ can be simplified away, so that we obtain a proof

$$x^{--} \cdot e \leftarrow_{R_3} \quad \overset{x^{--} \cdot (x^- \cdot x)}{\longrightarrow_{R_3} x.}$$

The initial equation $x^{--} \cdot e \approx x$ can now be deduced immediately (and then oriented). The set of all derived rules

$$\begin{aligned} e \cdot x &\rightarrow x \\ x^- \cdot x &\rightarrow e \\ (x \cdot y) \cdot z &\rightarrow x \cdot (y \cdot z) \\ x^- \cdot (x \cdot y) &\rightarrow y \\ x^{--} \cdot e &\rightarrow x \end{aligned}$$

is not convergent, however. For instance, the equation $e^- \cdot y \approx y$ is provable, but not by a rewrite proof. If we continue the completion process (and intersperse it with additional simplification rules) a convergent system of ten rules can be obtained eventually:

$e \cdot x$	$\rightarrow$	x		$x \cdot e$	$\rightarrow$	x
$x^- \cdot x$	$\rightarrow$	e		$x \cdot x^-$	$\rightarrow$	e
$(x \cdot y) \cdot z$	$\rightarrow$	$x \cdot (y \cdot z)$		x^{--}	$\rightarrow$	x
e^-	$\rightarrow$	e		$(x \cdot y)^-$	$\rightarrow$	$y^- \cdot x^-$
$x^- \cdot (x \cdot y)$	$\rightarrow$	y		$x \cdot (x^- \cdot y)$	$\rightarrow$	y

(Knuth and Bendix 1970).

2.3. Proof Simplification

Each proof transformation step by $\mathcal{R}_B$ decreases the complexity of a proof with respect to some well-founded measure. Consequently, there are no infinite proof transformation sequences.

Lemma 2.8. *The proof transformation system $\mathcal{R}_B$ is terminating.*

Proof. We define a measure of the complexity of a proof by assigning a certain cost to each proof step. The cost of a proof step $s \leftrightarrow^p_{u \approx_T v} t$ is the multiset $\{s, t\}$; the cost of a proof step $s \leftrightarrow^p_{u \approx_\perp v} t$ is the multiset $\{s\}$, if $u \succ v$, and $\{t\}$, if $v \succ u$. The complexity of a proof is the multiset of all costs of its proof steps. (Note that a proof P has the same complexity as the inverse proof P^{-1}.)

For example, the proof

$$x^{--} \cdot e \leftarrow_{R_2} x^{--} \cdot (x^- \cdot x) \leftrightarrow_{E_2} e \cdot x \rightarrow_{R_2} x$$

has complexity

$$\{\{x^{--} \cdot (x^- \cdot x)\}, \{x^{--} \cdot (x^- \cdot x), e \cdot x\}, \{e \cdot x\}\}.$$

We define an ordering $\succ_B$ on proofs by comparing proofs according to their complexity, using the twofold multiset extension $(\succ_{mul})_{mul}$ of the given reduction ordering $\succ$. The ordering $\succ_B$ is a proof reduction ordering and contains the transformation system $\mathcal{R}_B$. As a consequence, the transformation relation $\Rightarrow^+_B$ is well-founded. In detail:

i) *Orientation.* We have $s \leftrightarrow^\lambda_{s \approx_T t} t \succ_B s \rightarrow^\lambda_{s \approx_\perp t} t$, because $\{s, t\} \succ_{mul} \{s\}$.

ii) *Simplification.* We have $s \leftrightarrow^\lambda_{s \approx_T t} t \succ_B s \rightarrow^p_{s' \approx_\perp u'} u \leftrightarrow^\lambda_{u \approx_T t} t$, because $s \succ u$ and therefore $\{s, t\} \succ_{mul} \{s\}$ and $\{s, t\} \succ_{mul} \{u, t\}$.

iii) *Deletion.* Trivially, $s \leftrightarrow^\lambda_{s \approx_T s} s \succ_B \square$

iv) *Deduction.* We have $s \leftarrow^p_{w \approx_\perp v} u \rightarrow^q_{v' \approx_\perp w'} t \succ_B s \leftrightarrow^\lambda_{s \approx_T t} t$, because $\{u\} \succ_{mul} \{s, t\}$. Q.E.D.

Let $E_0 ; R_0 \vdash_\mathcal{B} E_1 ; R_1 \vdash_\mathcal{B} \cdots$ be a derivation in $\mathcal{B}$. Since the proof transformation system $\mathcal{R}_\mathcal{B}$ is terminating and reflects the inference system $\mathcal{B}$, every proof P_i in $E_i \cup R_i$ can be transformed by $\mathcal{R}_\mathcal{B}$ to a normal-form proof in $E_\infty \cup R_\infty$. To prove correctness of completion it suffices to show that these normal-form proofs are rewrite proofs. In other words, completion can be interpreted as a process of proof normalization, the goal of which is the derivation of rewrite proofs.

2.4. Fairness and Correctness

A large class of correct completion procedures can be characterized via the following notion of fairness.

Definition 2.9. Let $\mathcal{N}$ be a set of proofs, $\mathcal{R}$ be a proof transformation system, and $\mathcal{I}$ be an inference system. A derivation $E_0 \vdash_\mathcal{I} E_1 \vdash_\mathcal{I} \cdots$ is said to be *fair with respect to* $\mathcal{N}$ *and* $\mathcal{R}$ if for every proof P in E_∞ that is not contained in $\mathcal{N}$, there exists some proof Q in $\bigcup_i E_i$, such that $P \Rightarrow_\mathcal{R}^+ Q$.

Theorem 2.10. *Suppose $\mathcal{R}$ is a terminating proof transformation system and reflects the inference system $\mathcal{I}$. If a derivation $E_0 \vdash_\mathcal{I} E_1 \vdash_\mathcal{I} \cdots$ is fair with respect to $\mathcal{N}$ and $\mathcal{R}$, then every proof P in $\bigcup_i E_i$ can be transformed by $\mathcal{R}$ to a proof in $\mathcal{N}$.*

Proof. Suppose some proof in $\bigcup_i E_i$ cannot be transformed to a proof in $\mathcal{N}$. Let P be a minimal such proof with respect to the well-founded ordering $\Rightarrow_\mathcal{R}^+$. Since any proof in $\bigcup_i E_i$ can be transformed to a proof in E_∞, we may infer that P is a proof in E_∞. Since P is not contained in $\mathcal{N}$, we may use fairness to infer that there exists a proof Q in $\bigcup_i E_i$ with $P \Rightarrow_\mathcal{R}^+ Q$. This contradicts the assumption that P is in normal form with respect to $\mathcal{R}$. Q.E.D.

Let $\mathcal{N}_\mathcal{B}^\succ$ (or $\mathcal{N}_\mathcal{B}$) be the set of all rewrite proofs $s \rightarrow_R^* u \leftarrow_R^* t$, where R denotes any rewrite system contained in $\succ$. We shall derive sufficient conditions for the fairness (with respect to $\mathcal{N}_\mathcal{B}$ and $\mathcal{R}_\mathcal{B}$) of derivations in basic completion.

A proof in R is a rewrite proof if and only if it contains no peak $s \leftarrow_R u \rightarrow_R t$. In other words, fairness (in the case of basic completion) requires elimination of peaks. First observe that a peak can always be transformed to a rewrite proof, provided a rewrite proof exists.

Lemma 2.11. *If P is a peak $s \leftarrow_R u \rightarrow_R t$ and Q is a rewrite proof $s \rightarrow_R^* v \leftarrow_R^* t$, then $P \Rightarrow_\mathcal{B}^+ Q$.*

Proof. Transform the peak $s \leftarrow_R u \rightarrow_R t$ into an equality step $s \leftrightarrow_E t$ and then repeatedly simplify to obtain a proof $s \rightarrow_R^* v \leftrightarrow_E v \leftarrow_R^* t$ from which the equality step can be deleted. The result is a rewrite proof $s \rightarrow_R^* v \leftarrow_R^* t$ in R. Q.E.D.

If $s \leftarrow_R u \rightarrow_R t$ is a peak, but there exists no rewrite proof of $s \approx t$ in R, then new equations need to be deduced with which a rewrite proof can possibly be constructed. Certain equational consequences called "critical pairs" are of importance in this regard.

Definition 2.12. Let $s \approx t$ and $u \approx v$ be equations with no variables in common (if necessary the variables of one equation are renamed) and suppose that some non-variable subterm $s|_p$ of s is unifiable with u, σ being the most general unifier. Then the *superposition* of $u \approx v$ on $s \approx t$ at position p determines a *critical pair* $t\sigma \approx s\sigma[v\sigma]_p$. The proof $t\sigma \leftarrow_{t \approx s}^{\lambda} s\sigma \rightarrow_{u \approx v}^p s\sigma[v\sigma]_p$ is called a *critical overlap*; the term $s\sigma$, the *overlapped term*; the position p, the *critical pair position*.

For example, superposing the equation $x \cdot x^- \approx e$ on $(x \cdot y)^- \approx y^- \cdot x^-$ yields a critical pair $x^{--} \cdot x^- \approx e^-$. The corresponding critical overlap is $x^{--} \cdot x^- \leftarrow_E (x \cdot x^-)^- \rightarrow_E e^-$.

Critical pairs $t \approx t$ are called *trivial*. (For instance, critical pairs obtained by superposing an equation on literally the same equation at the topmost position $p = \lambda$ are trivial.) By $CP(E)$ we denote the set of all non-trivial critical pairs between equations in E. The set $CP(E)$ is finite whenever E is finite. Observe that usually $CP(E) \neq CP(E^{-1})$.

Lemma 2.13. (Critical Pair Lemma, Knuth and Bendix 1970) *If P is a peak $s \leftarrow_E u \rightarrow_E t$, then either there exists a rewrite proof $s \rightarrow_E^* v \leftarrow_E^* t$ or a "critical-pair proof" $s \leftrightarrow_{CP(E)} t$.*

Proof. Let E be a set of equations $\{v \approx w, v' \approx w'\}$ and P be a peak $s \leftarrow_{w \approx v}^p u \rightarrow_{v' \approx w'}^q t$ in E. If $s = t$, then the empty proof is a suitable rewrite proof of $s \approx t$. Let us therefore assume that s and t are distinct. Let σ and σ' be substitutions such that $u|_p = v\sigma$, $u|_q = v'\sigma'$, $s = u[w\sigma]_p$, and $t = u[w'\sigma']_q$. We distinguish three types of peaks, called non-overlaps, variable overlaps, and proper overlaps, respectively.

If the two positions p and q are disjoint, we speak of a *non-overlap*. Then $s|_q = (u[w\sigma]_p)|_q = u|_q = v'\sigma'$ and $t|_p = (u[w'\sigma']_q)|_p = u|_p = v\sigma$, so that there exists a rewrite proof $s \rightarrow_{v' \approx w'}^q u[w\sigma, w'\sigma']_{p,q} \leftarrow_{w \approx v}^p t$ (see Figure 2.2).

For example, since the two rules $x \cdot x^- \rightarrow e$ and $e \cdot x \rightarrow x$ are applied at disjoint positions in

$$e \cdot (e \cdot x) \leftarrow_R (x \cdot x^-) \cdot (e \cdot x) \rightarrow_R (x \cdot x^-) \cdot x$$

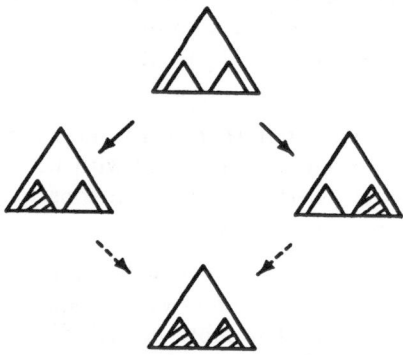

Figure 2.2: *Non-overlap*

the peak can be replaced by a rewrite proof

$$e \cdot (e \cdot x) \to_R e \cdot x \leftarrow_R (x \cdot x^-) \cdot x.$$

If p and q are not disjoint, one position has to be below the other. Let us assume, without loss of generality, that q is below p and let p' be a position such that $q = pp'$. Then $v\sigma|_{p'} = (u|_p)|_{p'} = u|_{pp'} = u|_q = v'\sigma'$. If p' is a position in v and $v|_{p'}$ is not a variable, we speak of a *proper overlap*. Otherwise, we speak of a *variable overlap* (or say that one proof step applies in the "variable part" of the other).

In other words, if P is a variable overlap, then there exist positions q' and q'', such that $p' = q'q''$ and $v|_{q'} = x$, for some variable x, and $v\sigma|_{p'} = v\sigma|_{q'q''} = (v\sigma|_{q'})|_{q''} = x\sigma|_{q''} = v'\sigma'$. Define the substitution τ so that $x\tau = x\sigma[w'\sigma']_{q''}$ and $y\tau = y\sigma$, for all variables y different from x. Since $x\sigma \to^{q''}_{v' \approx w'} x\tau$, there exists a proof $s|_p = w\sigma \to^*_{E'} w\tau$, where $E' = \{v' \approx w'\}$. Let now $p_1, \ldots, p_n$ be all those (pairwise disjoint) positions in v at which the variable x occurs. Since $t|_p = (u[w'\sigma']_q)|_p = (u[w'\sigma']_{pp'})|_p = u|_p[w'\sigma']_{p'} = v\sigma[w'\sigma']_{p'} = v\sigma[x\tau]_{q'}$ and q' is one of the positions p_i, there is a proof $t|_p \to_{E'} v\sigma[x\tau, \ldots, x\tau]_{p_1, \ldots, p_n} = v\tau$. In sum, we have $s = u[w\sigma]_p \to^*_{E'} u[w\tau]_p$ and $t \to_{E'} u[v\tau]_p$. Since $E' \subseteq E$, we may conclude that there exists a rewrite proof $s \to^*_E s' \leftarrow_E t' \leftarrow^*_E t$ (cf. Figure 2.3).

Suppose, for example, that R contains the rules $x \cdot x^- \to e$ and $e \cdot x \to x$. The variable overlap

$$e \leftarrow_R (e \cdot x) \cdot (e \cdot x)^- \to_R x \cdot (e \cdot x)^-$$

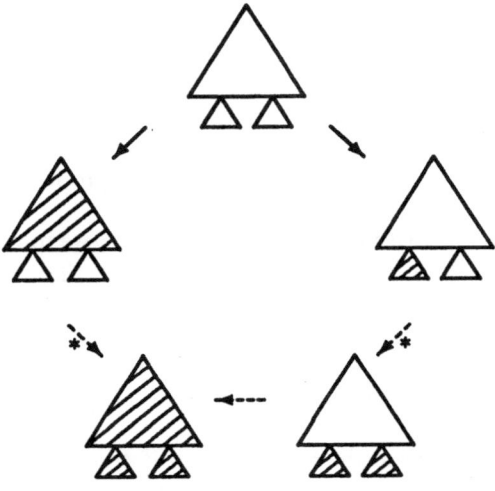

Figure 2.3: *Variable overlap*

can be replaced by a rewrite proof

$$e \leftarrow_R x \cdot x^- \leftarrow_R x \cdot (e \cdot x)^-.$$

Finally, suppose P is a proper overlap. Let us assume without loss of generality that $v \approx w$ and $v' \approx w'$ have no variables in common. (If the two equations do have variables in common, consider the peak $s \leftarrow_{w \approx v}^p u \rightarrow_{v'' \approx w''}^q t$ instead, where $v'' \approx w''$ is literally similar to $v' \approx w'$, but has no variables in common with $v \approx w$.) Since $(v|_{p'})\sigma = v\sigma|_{p'} = v'\sigma'$, the two terms $v|_{p'}$ and v' are unifiable. Let τ be a most general unifier and let τ' be a substitution, such that $(x\tau)\tau' = x\sigma$, for all variables x occurring in $v \approx w$ or $v' \approx w'$. Then $P = u[Q\tau']_p$, where Q is the critical overlap $w\tau \leftarrow_{w \approx v}^\lambda v\tau \rightarrow_{v' \approx w'}^{p'} v\tau[w'\tau]_{p'}$. Q.E.D.

Note that the proper-overlap case can be characterized more precisely:

Corollary 2.14. *If P is a proper overlap $s \leftarrow_{w \approx v}^p u \rightarrow_{v' \approx w'}^q t$, where $v \approx w$ and $v' \approx w'$ have no variables in common, then P can be written as $u[Q\tau]_r$, for some term u, position r in u, substitution τ, and critical overlap Q.*

The Critical Pair Lemma suggests the following sufficient condition for fairness.

Lemma 2.15. *A non-failing derivation* $E_0 ; R_0 \vdash_{\mathcal{B}} E_1 ; R_1 \vdash_{\mathcal{B}} \cdots$ *in basic completion is fair (with respect to* $\mathcal{N}_{\mathcal{B}}$ *and* $\mathcal{R}_{\mathcal{B}}$*) if the set of critical pairs* $CP(R_\infty)$ *is a subset of the set* $\bigcup_k E_k$ *of all deduced equations.*

Proof. Let $E_0 ; R_0 \vdash_{\mathcal{B}} E_1 ; R_1 \vdash_{\mathcal{B}} \cdots$ be a derivation for which $E_\infty = \emptyset$ and $CP(R_\infty)$ is a subset of $\bigcup_k E_k$. It suffices to prove that for every peak P of the form $s \leftarrow_{R_\infty} u \rightarrow_{R_\infty} t$, there exists a proof Q in $\bigcup_k (E_k \cup R_k)$, such that $P \Rightarrow_{\mathcal{B}}^+ Q$.

If there exists a rewrite proof $s \rightarrow_{R_\infty}^* v \leftarrow_{R_\infty}^* t$, we may use Lemma 2.11 to infer that $P \Rightarrow_{\mathcal{B}}^+ Q$, for some proof Q in R_∞. If there exists no such rewrite proof, then by the Critical Pair Lemma (and Corollary 2.14) $P = w[P'\tau']_p$, for some term w, position p in w, substitution τ', and critical overlap P'. But then $s \leftrightarrow_{CP(R_\infty)} t$ so that by fairness $s \leftrightarrow_{E_j} t$, for some $j \geq 0$. This implies that there is a suitable proof Q with $P \Rightarrow_{\mathcal{B}}^+ Q$. Q.E.D.

Usually, a completion procedure is called fair if it generates only derivations for which all critical pairs between persisting rules are computed.[1]

Applying Theorem 2.10 to basic completion we obtain:

Proposition 2.16. (Correctness) *If a non-failing derivation in basic completion is fair, then the limit rewrite system* R_∞ *is convergent.*

Proof. Lemmas 2.7, 2.8, and 2.15 show that Theorem 2.10 can be applied to the basic completion system $\mathcal{B}^\succ$, the set of proofs $\mathcal{N}_{\mathcal{B}}$, and the transformation system $\mathcal{R}_{\mathcal{B}}$. Consider a non-failing derivation in basic completion. The limit rewrite system R_∞ is contained in $\succ$ and hence is terminating. Since the derivation does not fail, we have $E_\infty = \emptyset$. Also, by Theorem 2.10 every proof P in R_∞ can be transformed to a rewrite proof, which implies that R_∞ is Church-Rosser. In sum, R_∞ is convergent. Q.E.D.

The well-known characterization of the Church-Rosser property of terminating rewrite systems in terms of critical pairs can also be easily proved via proof orderings.

Definition 2.17. A rewrite system R is *convergent on a set of equations* E if every equation $s \approx t$ in E is provable by a rewrite proof $s \rightarrow_R^* v \leftarrow_R^* t$.

Proposition 2.18. *A terminating rewrite system R is Church-Rosser if and only if it is convergent on the set of critical pairs $CP(R)$.*

---

[1] Dershowitz (1989) insists on a stronger fairness condition which requires a procedure to produce a fair derivation whenever one exists. Fairness in this stronger sense may require the enumeration of all possible derivations from a given initial set $E_0 \cup R_0$.

Proof. The only-if direction is obvious. For the other direction, let $\succ$ be the reduction ordering $\rightarrow_R^+$ (note that R is terminating) and $\succ_B$ be the corresponding proof ordering. We claim that whenever P is a minimal proof (with respect to $\succ_B$) of an equation $s \approx t$ in R, then it is a rewrite proof. For if P is not a rewrite proof, it must contain a peak $s' \leftarrow_R u' \rightarrow_R t'$. But then we may use the Critical Pair Lemma in combination with the fact that R is convergent on $CP(R)$, to infer that there is a rewrite proof $s' \rightarrow_R^* v' \leftarrow_R^* t'$, which contradicts the minimality of P. In sum, if an equation is provable, then it is provable by a rewrite proof. This implies that R is Church-Rosser. Q.E.D.

A rewrite system R is said to be *non-overlapping* if no left-hand side s of a rule in R is unifiable with any term that is literally similar to a non-variable subterm of some other left-hand side or a proper non-variable subterm of s. If R is non-overlapping, then $CP(R) = \emptyset$. Non-overlapping, terminating systems are therefore convergent.

Typical examples of non-overlapping systems are many equational programs, such as the following rewrite system for *symbolic differentiation* ("with respect to X"):

$$
\begin{array}{rcl}
D(\mathrm{X}) & \rightarrow & 1 \\
D(\mathrm{Y}) & \rightarrow & 0 \\
D(-x) & \rightarrow & -D(x) \\
D(x + y) & \rightarrow & D(x) + D(y) \\
D(x \times y) & \rightarrow & D(x) \times y + x \times D(y) \\
D(x \div y) & \rightarrow & D(x) \div y - (x \times D(y)) \div (y^2) \\
D(\ln x) & \rightarrow & D(x) \div x \\
D(x^y) & \rightarrow & D(x) \times (y \times x^{y-1}) + ((\ln x) \times D(y)) \times (x^y)
\end{array}
$$

(Knuth 1973, p. 337). This rewrite system is non-overlapping and terminating; hence convergent. (Termination can be proved by a recursive path ordering with a total precedence in which D is maximal.)

Having outlined our approach by using the basic concepts of inference systems, proof orderings, and normal-form proofs to prove the correctness of basic completion, we shall proceed with a discussion of more sophisticated completion methods.

2.5. Standard Completion

Basic completion consists of fundamental inference rules commonly used in constructing convergent systems. For efficiency reasons additional features, such as more extensive simplification, are indispensable in practice.

The following inference rules suffice to formulate such simplification mechanisms:

COMPOSITION: $$\frac{E\,;R\cup\{s\to t\}}{E\,;R\cup\{s\to u\}}\quad \text{if } t\to_R u$$

COLLAPSE: $$\frac{E\,;R\cup\{s\to t\}}{E\cup\{u\approx t\}\,;R}\quad \begin{array}{l}\text{if there is a proof } s\to^p_{v\approx_\perp w} u\\ \text{in } R, \text{ where } s\,\triangleright\,v\end{array}$$

The symbol $\triangleright$ denotes the strict part of the *encompassment ordering*,[2] which is defined as follows: $s\,\underset{\sim}{\triangleright}\,t$ if some subterm of s is an instance of t, but not vice versa. For example, $x+(x+0)\,\triangleright\,x+y$.

By a *standard completion system* we mean any inference system $C^\curvearrowright$ consisting of $B^\curvearrowright$ plus the inference rules above.

Composition allows simplification of right-hand sides of rewrite rules; collapse, of left-hand sides. While composition produces another rewrite rule, the equation obtained by collapsing a rule need not necessarily be orientable with respect to the given reduction ordering $\succ$. Note that a rewrite rule must not be used to collapse another rewrite rule with a literally similar left-hand side. This restriction is of no relevance in practice, for if equations are always fully simplified before they are oriented (as is the case in most implementations of completion), then no two rules with literally similar left-hand sides can be produced. In such procedures the collapse rule can be applied without explicitly checking whether the condition $s\,\triangleright\,v$ is satisfied.

Inference systems $C^\curvearrowright$ are evidently sound, in that the congruence relations $\leftrightarrow^*_{E\cup R}$ and $\leftrightarrow^*_{E'\cup R'}$ are the same whenever $E\,;R\vdash_C E'\,;R'$. Also, if $E\,;R\vdash_C E'\,;R'$ and the reduction ordering $\succ$ contains R, then $\succ$ also contains R'. Consequently, the system R_∞ is terminating for any derivation for which the initial rewrite system R_0 is contained in the reduction ordering $\succ$. As with basic completion, the inference rules never decrease the strength of rewriting: if $E\,;R\vdash_C E'\,;R'$ then any term t that is reducible by R is also reducible by R'.

Collapse and composition are reflected by proof transformation rules of the form

$$s\to^\lambda_{s\approx_\perp t} t\quad\Rightarrow\quad s\to^\lambda_{s\approx_\perp u} u\leftarrow^p_{u'\approx_\perp t'} t$$
$$s\to^\lambda_{s\approx_\perp t} t\quad\Rightarrow\quad s\to^q_{v\approx_\perp w} u\leftrightarrow^\lambda_{u\approx_\top t} t$$

where $s\succ t$, $s\succ u$, $t'\succ u'$, $v\succ w$, and $s\,\triangleright\,v$ (see Figure 2.4).

---

[2] The collapse rule can be defined in terms of an arbitrary well-founded ordering, but virtually all implementations of completion are based on the encompassment ordering.

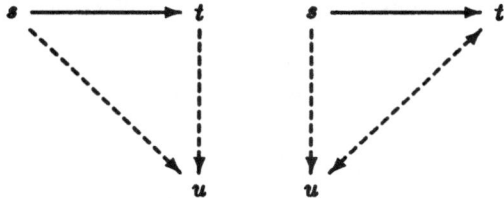

Figure 2.4: *Transformation rules for standard completion*

We denote by $\mathcal{R}_{\mathcal{C}}^{\succ}$ (or simply $\mathcal{R}_{\mathcal{C}}$) the set consisting of $\mathcal{R}_{\mathcal{B}}^{\succ}$ plus all these transformation rules. The corresponding transformation relation is denoted by $\Rightarrow_{\mathcal{C}}$.

With these definitions we have:

Lemma 2.19. *The proof transformation system $\mathcal{R}_{\mathcal{C}}$ reflects the standard completion system C.*

Lemma 2.20. *The transformation system $\mathcal{R}_{\mathcal{C}}$ is terminating.*

Proof. We redefine the cost assigned to single proof steps to accommodate composition and collapse. The cost of a proof step $s \leftrightarrow^{p}_{u \approx_{T} v} t$ is the quadruple $(\{s,t\}, \perp, \perp, \perp)$; the cost of a proof step $s \leftrightarrow^{p}_{u \approx_{\perp} v} t$ is the quadruple $(\{s\}, s|_{p}, u, t)$, if $u \succ v$, and $(\{t\}, t|_{p}, v, s)$, if $v \succ u$. Again, the complexity of a proof is the multiset of all costs of its proof steps. (Thus a proof P has the same complexity as the inverse proof P^{-1}.)

Proof steps are compared according to their complexity using the lexicographic combination $\succ^{c}$ of the multiset extension $\succ_{mul}$ of the reduction ordering $\succ$, the subterm ordering, the encompassment ordering $\unrhd$, and the reduction ordering $\succ$. To compare proofs we use the multiset extension of this ordering. The corresponding ordering $\succ_{\mathcal{C}}$ is a proof reduction ordering and contains the proof transformation system $\mathcal{R}_{\mathcal{C}}$. This can easily be verified for the proof transformation rules of $\mathcal{R}_{\mathcal{B}}$, using the same arguments as in Lemma 2.8. In the remaining cases we have:

i) *Composition.* We have $s \rightarrow^{\lambda}_{s \approx_{\perp} t} t \succ_{\mathcal{C}} s \rightarrow^{\lambda}_{s \approx_{\perp} u} u \leftarrow^{p}_{u' \approx_{\perp} t'} t$, as $s \succ t$ and $t' \succ u'$ imply

$$\{(\{s\}, s, s, t)\} \succ_{\mathcal{C}} \{(\{s\}, s, s, u), (\{t\}, t|_{p}, t', u)\}.$$

ii) *Collapse.* Since $s \succ t$, $s \succ u$, and $s \unrhd s'$, we have

$$\{(\{s\}, s, s, t)\} \succ_{\mathcal{C}} \{(\{s\}, s|_{q}, v, u), (\{t,u\}, \perp, \perp, \perp)\},$$

which implies $s \rightarrow^{\lambda}_{s \approx_{\perp} t} t \succ_{\mathcal{C}} s \rightarrow^{q}_{v \approx_{\perp} w} u \leftrightarrow^{\lambda}_{u \approx_{T} t} t$.   Q.E.D.

The following theorem is an immediate corollary of Theorem 2.10:

Theorem 2.21. *If a non-failing derivation by standard completion procedure is fair (with respect to $\mathcal{N}_B$ and $\mathcal{R}_C$), then the limit rewrite system R_∞ is convergent.*

It can easily be shown that a derivation is fair if all critical pairs between persisting rules are computed, i.e., if $CP(R_\infty)$ is a subset of $\bigcup_k E_k$. For the purpose of establishing the correctness of most practical completion procedures the following stronger result is essential.

Proposition 2.22. *A non-failing derivation $E_0 \,; R_0 \vdash_C E_1 \,; R_1 \vdash_C \cdots$ is fair (with respect to $\mathcal{N}_B$ and $\mathcal{R}_C$) if for every critical overlap P of the form $s \leftarrow_{R_\infty} u \rightarrow_{R_\infty} t$ there exists a proof Q of $s \approx t$ in $\bigcup_i (E_i \cup R_i)$, such that $P \succ_C Q$.*

Proof. We will prove that every proof of $s \approx t$ in $\bigcup_i (E_i \cup R_i)$ can be transformed by $\mathcal{R}_C$ to a rewrite proof in R_∞. Suppose some proof cannot be transformed to a rewrite proof in R_∞. Let P be one such proof that is minimal with respect to the proof ordering $\succ_C$.

Since every proof can be transformed by $\mathcal{R}_C$ to a proof in $E_\infty \cup R_\infty$ and $E_\infty = \emptyset$, we may infer that P is a proof in R_∞, but contains a peak $s' \leftarrow_{R_\infty} u' \rightarrow_{R_\infty} t'$. If this peak is not a proper overlap, then there exists a rewrite proof $s' \rightarrow_{R_\infty} v' \leftarrow_{R_\infty} t'$ and we may use Lemma 2.11 to infer that there is a proof Q such that $P \Rightarrow_C^+ Q$. Since $P \succ_C Q$, the proof Q can be transformed to a rewrite proof in R_∞, which is a contradiction.

If the peak is a proper overlap, then it can be written as $u[Q''\sigma]_p$, where Q'' is a critical overlap between rules in R_∞. By assumption there is a proof Q' of $s' \approx t'$ such that $Q'' \succ_C Q'$, and consequently $u[Q''\sigma]_p \succ_C u[Q'\sigma]_p$. Since $P \succ_C u[Q'\sigma]_p$, the proof $u[Q'\sigma]_p$ can be transformed to a rewrite proof $s' \rightarrow_{R_\infty} v' \leftarrow_{R_\infty} t'$. But then there is a proof Q such that $P \Rightarrow_C^+ Q$, which is a contradiction.   Q.E.D.

The notion of completion, as we have formalized it above, covers a wide variety of specific completion procedures, including the procedures described by Huet (1981) and Dershowitz (1982b). The latter procedure is described below. It accepts as input a set of equations E_0 and a reduction ordering $\succ$.

Let E be E_0 and R be the empty set. Then repeat as long as equations are left in E. If none remain, terminate successfully.

1. Remove an equation $s \approx t$ (or $t \approx s$) from E such that $s \succ t$. If none exists, terminate with failure.

2. Add the rule $s \rightarrow t$ to R.

3. Use R to reduce the right-hand sides of existing rules.

4. Add to E all critical pairs formed using the new rule.

5. (Optional) Remove all old rules from R whose left-hand side contains an instance of s.

6. Use R to reduce both sides of equations in E to normal forms. Remove any equation whose reduced sides are identical.

This procedure is a standard completion procedure. Step 2 represents an application of orientation; step 3, repeated application of composition; step 4, repeated deduction; step 6, repeated simplification and deletion. Step 4, in combination with step 5, implicitly uses collapse. For whenever a rule $s \rightarrow t$ can be collapsed to $u \approx t$, then $u \approx t$ is a critical pair in $CP(R)$. In the above procedure, the equation $u \approx t$ is first deduced in step 4, but may be deleted in step 5.

Observe that equations in E are kept in fully simplified form (step 6). Consequently, whenever an equation $s \approx t$ is selected in step 1, then both s and t are irreducible with respect to the current rewrite system R. This guarantees that R will never contain two rules with literally similar left-hand sides, so that the condition for applying collapse inferences is always satisfied.

The above procedure may generate critical pairs from non-persisting rules that are not necessary. On the other hand, not all critical pairs between persisting rules are generated either. For if the right-hand side of a rewrite rule is simplified, in step 3, no critical pairs are computed with the new simplified rule. Nonetheless, according to Proposition 2.22 the procedure is correct.

More precisely, suppose there is a proof $t \rightarrow_R t'$ and some rule $s \rightarrow t$ in R is replaced by $s \rightarrow t'$ as a result of composition. Then for every critical overlap P involving the new rule $s \rightarrow t'$, there exists a proof Q with $P \succ_C Q$. Suppose P is a critical overlap $v\sigma \leftarrow_{v \approx u}^{\lambda} u\sigma[s\sigma]_p \rightarrow_{s \approx t'}^{p} u\sigma[t'\sigma]_p$. Since the critical pair $v\sigma \approx u\sigma[t\sigma]_p$, which is obtained from the overlap $v\sigma \leftarrow_{v \approx u}^{\lambda} u\sigma[s\sigma]_p \rightarrow_{s \approx t}^{p} u\sigma[t\sigma]_p$, has already been computed, there exists a proof $Q = v\sigma \leftrightarrow_{v\sigma \approx u\sigma[t\sigma]_p}^{\lambda} u\sigma[t\sigma]_p \leftarrow_R u\sigma[t'\sigma]$ in which all terms are strictly smaller than $u\sigma$. As a consequence, we have $P \succ_C Q$. A similar argument applies to critical overlaps $t'\sigma \leftarrow_{t' \approx s}^{\lambda} s\sigma[u\sigma]_p \rightarrow_{u \approx v}^{p} s\sigma[v\sigma]_p$.

The extensive use of simplification, composition, and collapse is typical of completion procedures.

Definition 2.23. A derivation is called *simplifying* if R_∞ is a reduced rewrite system and all equations in E_∞ are unorientable with respect to $\succ$, yet irreducible by R_∞.

In other words, if a derivation is simplifying then all persisting equations are fully simplified and the persisting rules form a reduced rewrite system. Most implementations of standard completion procedures are simplifying in this sense: they construct reduced convergent rewrite systems, whenever they succeed.

Implementations of standard completion have been described, among others, by Lescanne (1983) and Kapur and Sivakumar (1984). Lescanne (1989) describes an implementation of completion that is directly based on inference rules (and a suitable control).

Completion has been applied to a variety of problems including the word problem in universal and finitely presented algebras (Knuth and Bendix 1970, Le Chenadec 1986) and equational programming (Dershowitz 1985). A number of convergent system have been derived with such procedures. The following examples are from Knuth and Bendix (1970).

Example 2.24. The theory of a single axiom $x^- \cdot (x \cdot y) \approx y$, expressing a property of the *inverse* operator, can be represented as a convergent system of three rules:

$$
\begin{aligned}
x^- \cdot (x \cdot y) &\to y \\
x^{--} \cdot y &\to x \cdot y \\
x \cdot (x^- \cdot y) &\to y
\end{aligned}
$$

Example 2.25. *Central groupoids* are also characterized by a single axiom $(x \cdot y) \cdot (y \cdot z) \approx y$ and can be represented as a convergent system of three rules:

$$
\begin{aligned}
(x \cdot y) \cdot (y \cdot z) &\to y \\
x \cdot ((x \cdot y) \cdot z) &\to x \cdot y \\
(x \cdot (y \cdot z)) \cdot z &\to y \cdot z
\end{aligned}
$$

Example 2.26. Algebraic systems with a left identity and a right inverse are called *left groups*:

$$
\begin{aligned}
e \cdot x &\approx x \\
x \cdot x^- &\approx e \\
(x \cdot y) \cdot z &\approx x \cdot (y \cdot z)
\end{aligned}
$$

A convergent rewrite system for this theory is:

$$
\begin{aligned}
e \cdot x &\to x & e^- &\to e \\
x \cdot x^- &\to e & x \cdot e &\to x^{--} \\
(x \cdot y) \cdot z &\to x \cdot (y \cdot z) & x^{--} \cdot y &\to x \cdot y \\
x \cdot (x^- \cdot y) &\to y & x^- \cdot (x \cdot y) &\to y \\
x^{---} &\to x^- & (x \cdot y)^- &\to y^- \cdot x^-
\end{aligned}
$$

The correctness of a specific completion procedure was first proved by Huet (1981). An advantage of formulating completion as an equational inference system is that correctness can be proved for a wide class of different completion procedures. The proof ordering approach is comparatively simple and intuitive, especially for dealing with simplification inferences. The intrinsic difficulty of applying the approach consists in finding a suitable ordering to establish proof normalization. Once an appropriate proof ordering has been found, the remaining verification steps are straightforward.

Various minor improvements of standard completion can readily be verified in the proof ordering framework. For instance, since fairness requires only computation of critical pairs between persisting rules, critical pairs that originated from a collapsed rule can be deleted (provided, of course, that they have not yet been used for simplification). Furthermore, an inference rule,

$$\text{SUBSUMPTION:} \quad \frac{E \cup \{s \approx t, u[s\sigma]_p \simeq u[t\sigma]_p\} \, ; R}{E \cup \{s \approx t\} \, ; R}$$

can be added to standard completion without impairing correctness. "Orientable instances" of equations can be used for simplification. That is, if $u \approx v$ is an equation and $u\sigma \succ v\sigma$, then the rewrite rule $u\sigma \to v\sigma$ can be used for simplification (see Chapter 4).

Inference rules can also be formulated in terms of a proof ordering. For example,

$$\text{DELETION:} \quad \frac{E \cup \{s \approx t\} \, ; R}{E \, ; R} \quad \text{if } s \leftrightarrow^\lambda_{s \approx t} t \succ_c P, \text{ for some proof } P \text{ in } E \cup R$$

is a generalized version of the deletion inference rule. The inference rules for simplification of rewrite rules, composition and collapse, can be generalized in a similar way. Further details can be found in Chapter 4. Similar inference rules have turned out to be particularly useful in the context of conditional completion procedures (Ganzinger 1987).

2.6. Critical Pair Criteria

The efficiency of the completion process depends primarily on the number of critical pairs that are generated. Simplification can be very effective in eliminating superfluous equations. For instance, any critical pair $s \approx t$ for which both s and t can be reduced to a common normal form, is redundant. But systematic normalization of terms can be costly, while the redundancy of a critical pair can often be determined more efficiently by examining the structure of the associated critical overlap.

Definition 2.27. A *critical pair criterion* is a mapping CPC on sets of (labelled) equations, such that $CPC(E) \subseteq CP(E)$.

Critical pairs in $CPC(E)$ are meant to be redundant.

Definition 2.28. A critical pair criterion CPC is said to be *sound* if, for all terminating rewrite systems R, whenever R is convergent on $CP(R) \setminus CPC(R)$, then it is convergent.

In other words, a sound criterion provides a characterization of the Church-Rosser property. To be of use in practice, a criterion also has to be compatible with the simplification mechanisms employed by completion.

Definition 2.29. A derivation $E_0 ; R_0 \vdash_C E_1 ; R_1 \vdash_C \cdots$ in standard completion is said to be *fair with respect to* a critical pair criterion CPC if for every critical overlap P associated with a critical pair $s \approx t$ in $CP(R_\infty) \setminus \bigcup_i CPC(E_i \cup R_i)$ there exists a proof Q of $s \approx t$ in $\bigcup_i (E_i \cup R_i)$, such that $P \succ_C Q$. A criterion CPC is *correct* if every non-failing derivation that is fair with respect to CPC is also fair with respect to $\mathcal{R}_C$ and $\mathcal{N}_B$.

Evidently, if a non-failing derivation is fair with respect to a correct criterion CPC, then R_∞ is canonical. The following lemma follows from Proposition 2.22.

Lemma 2.30. *A critical pair criterion CPC is correct if for every set of equations E and every critical overlap P associated with a critical pair $s \approx t$ in $CPC(E)$ there exists a proof Q of $s \approx t$ in E, such that $P \succ_C Q$.*

Proposition 2.31. *Any correct critical pair criterion is sound.*

Proof. Let CPC be a correct criterion, R be a terminating rewrite system, and $\succ$ be the reduction ordering induced by R. Furthermore, suppose R is convergent on $CP(R) \setminus CPC(R)$. We have to show that R is Church-Rosser. The (one-step) non-failing derivation $\emptyset ; R$ is fair with respect to CPC. By the correctness of CPC it is also fair with respect to $\mathcal{N}_B$ and $\mathcal{R}_C$. By Theorem 2.21 the rewrite system R is Church-Rosser. Q.E.D.

A sound criterion need not necessarily be correct. For instance, Zhang and Kapur (1989) suggest a criterion PCP for standard completion, where $PCP(E \cup R)$ consists of all critical pairs of a rule $u \to v$ in R on another rule $s \to t$ in R at a position pq, such that $p \neq \lambda \neq q$ and the overlapped term $s\sigma$ is reducible at position p. This criterion can be shown to be sound without much difficulty, but is not correct.[3] For instance, the derivation

$$\emptyset\,;\{h(a) \to a, f(g(x,h(x))) \to a, g(a,h(a)) \to a\}$$
$$\vdash_C \quad \{g(a,a) \approx a\}\,;\{h(a) \to a, f(g(x,h(x))) \to a\}$$
$$\vdash_C \quad \emptyset\,;\{h(a) \to a, f(g(x,h(x))) \to a, g(a,a) \to a\}$$

is fair with respect to PCP. The only critical pair between the final three rules is $a \approx f(g(a,a))$, which is obtained by superposition of $h(a) \to a$ on $f(g(x,h(x))) \to a$. This critical pair is in $PCP(E_0 \cup R_0)$ and therefore is superfluous according to the criterion. But even though the above derivation is fair with respect to PCP, the final rewrite system is not convergent, as the term $f(g(a,h(a)))$ has two different normal forms $f(a)$ and a.

An example of a correct criterion is *blocking*, a concept introduced by Slagle (1974) and applied to rewriting by Lankford and Ballantyne (1979).

Definition 2.32. Let E be a set of equations and $t\sigma \leftarrow^\lambda_{t\approx s}$, $s\sigma \to^p_{u\approx v}$ $s\sigma[v\sigma]_p$ be a critical overlap in E. The overlap is said to be *non-blocked* (with respect to $\succ$) if there exist a term w and a variable x in s or u, such that $x\sigma \to_E w$ and $x\sigma \succ w$. Otherwise, it is called *blocked*.

A critical pair is blocked if the corresponding critical overlap is blocked. The set $BCP(E)$ consists of all non-blocked critical pairs in $CP(E)$.

Proposition 2.33. *The non-blocked criterion BCP is correct.*

Proof. If $t\sigma \approx s\sigma[v\sigma]_p$ is a non-blocked critical pair, then $x\sigma \to_E w$, where $x\sigma \succ w$ and x occurs in s or u. Let σ' be the substitution for which $x\sigma' = w$ and $y\sigma' = y\sigma$, for all variables y distinct from x. Let Q be the proof $t\sigma \to^*_E t\sigma' \leftarrow_E s\sigma' \to_E s\sigma'[v\sigma']_p \leftarrow^*_E s\sigma[v\sigma]_p$ (see Figure 2.5). Then $P \succ_c Q$ and we may use Lemma 2.30 to infer that the criterion BCP is correct.   Q.E.D.

The proposition shows that non-blocked critical pairs may be ignored by completion. The non-blocked criterion BCP is a special case of the following connectedness criterion.

Definition 2.34. Let E be a set of equations. Two terms s and t are said to be *connected below u in E* (with respect to $\succ$) if there are terms $u_1,\ldots,u_n$, such that $s \leftrightarrow_E u_1 \leftrightarrow_E \cdots \leftrightarrow_E u_n \leftrightarrow_E t$ and $u \succ u_i$, for all i with $1 \leq i \leq n$.

---

[3] Zhang and Kapur attribute the criterion to Winkler and Buchberger (1983), but the latters' actual, more subtle criterion is a specific instance of the composite criterion described below, and is correct.

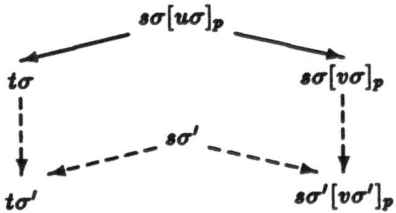

Figure 2.5: *Non-blocked overlap*

Connectedness was introduced by Buchberger (1984). The inference rules of standard completion preserve connectedness. That is, whenever $E\,;R \vdash_C E'\,;R'$ and two terms s and t are connected below u in $E \cup R$, then they are also connected below u in $E' \cup R'$. The concept can readily be utilized for a critical pair criterion, as we may view completion as a process of establishing, for every critical overlap $s \leftarrow_R u \rightarrow_R t$, the connectedness of s and t below the overlapped term u. For instance, adding the critical pair $s \approx t$ as a new equation is one way of establishing connectedness. Conversely, if s and t are already connected, then the critical pair $s \approx t$ is superfluous.

We define $CCP(E)$ as the set of all critical pairs $s \approx t$ in $CP(E)$ such that s and t are connected below the associated overlapped term u.

Proposition 2.35. *The connected critical pair criterion CCP is correct.*

Proof. Observe that if P is a critical overlap of the form $s \leftarrow_R u \rightarrow_R t$, where s and t are connected below the overlapped term u, then there exists a proof Q of $s \approx t$, such that $P \succ_C Q$.   Q.E.D.

A criterion based on connectedness was first formulated by Buchberger (1979) for a completion-like algorithm for constructing canonical bases for polynomial ideals. Various criteria for standard completion, based on connectedness with respect to the reduction ordering $\rightarrow_R^+$ induced by R, have been developed by Winkler and Buchberger (1983), Winkler (1985), and Küchlin (1985, 1986b). These criteria differ in the respective tests used to ensure connectedness. Let us sketch the basic idea.

Suppose $s \leftarrow_R u \rightarrow_R t$ is a critical overlap and u can be rewritten to some term v by R. We can decompose the original overlap into two peaks $s \leftarrow_R u \rightarrow_R v$ and $v \leftarrow_R u \rightarrow_R t$. If $s \leftarrow_R u \rightarrow_R v$ is a non-overlap or a variable overlap, then s and v are connected below u. Otherwise, the equation $s \approx v$ can be written as $u'[s'\sigma]_p \approx u'[v'\sigma]_p$, for some critical pair

$s' \approx v'$. If this critical pair has already been computed, then s and v are also connected below u. Similar arguments apply to v and t. Connectedness can therefore often be verified by checking whether certain critical pairs have already been computed. Book-keeping mechanisms to that end have been proposed by Küchlin (1985, 1986b) and Winkler (1985).[4]

For instance, suppose that all rewrite rules are numbered during completion in such a way that no two rules in any system R_n are assigned the same number. In addition, a rule is marked once all critical pairs with lower-numbered rules have been computed. Let P be a critical overlap as described above and suppose the rules numbered i, j, and k, respectively, are applied in the proof steps $u \rightarrow_R s$, $u \rightarrow_R t$, and $u \rightarrow_R v$. If rule $max(i, k)$ is marked, then all critical pairs between the corresponding rules have been computed, so that s and v are connected below u. Similarly, if rule $max(j, k)$ is marked, then t and v are connected below u.

The emphasis in the papers cited above is on soundness and practicality, though Winkler (1985) and Küchlin (1986b) also prove the correctness of specific versions of completion that incorporate tests for connectedness. Winkler's proof is similar to the proof of correctness of standard completion in Huet (1981); Küchlin's proof is based on multiset induction. Harald Ganzinger [personal communication] has pointed out that connectedness criteria are implicitly used in most completion procedures, e.g., in the procedures described by Huet (1981) and Dershowitz (1982b).

Connectedness is essentially a concept based on *terms*. A similar idea can be applied to equational *proofs*, which motivates the following definition.

Definition 2.36. A peak $s \leftarrow_E u \rightarrow_E t$ is said to be *composite* in E (with respect to the proof ordering $\succ_C$) if there are proofs $P_1, \ldots, P_n$ in E, where P_i is a proof of $u_{i-1} \approx u_i$, such that $u_0 = s$, $u_{n+1} = t$, $u \succ u_i$, and $(s \leftarrow_E u \rightarrow_E t) \succ_C P_i$. The proof $P_1 \cdots P_n$ is called a *decomposition* of the peak $s \leftarrow_E u \rightarrow_E t$.

A critical pair $s \approx t$ is *composite* if its corresponding critical overlap is composite. By $PCP(E)$ we denote the set of all composite critical pairs of E.

Proposition 2.37. *The composite criterion PCP is correct.*

Proof. Let $E_0 \,; R_0 \vdash_C E_1 \,; R_1 \vdash_C \cdots$ be a derivation that is fair with respect to PCP and for which $E_\infty = \emptyset$. We prove by induction on $\succ_C$ that

---

[4]The test described by Winkler restricts the position at which the rewrite step $u \rightarrow_R v$ may apply, whereas no such restriction is imposed by Küchlin.

every proof P in $\bigcup_i (E_i \cup R_i)$ can be transformed by $\Rightarrow_{\mathcal{C}}^{+}$ to a rewrite proof in R_∞. The assertion is assumed to be true for all proofs Q with $P \succ_{\mathcal{C}} Q$.

If P is not a proof in R_∞ or contains a non-proper overlap, then there is some proof Q, such that $P \Rightarrow_{\mathcal{C}}^{+} Q$. Suppose P is a proof in R_∞, but contains a proper overlap $s \leftarrow_{R_i} u \rightarrow_{R_i} t$. This proper overlap can be written as $u[P'\sigma]_p$, where P' is a critical overlap. Let $s' \approx t'$ be the corresponding critical pair in $CP(R_\infty)$. If the critical pair $s' \approx t'$ is contained in some set E_k, then there is some proof Q in $\bigcup_i (E_i \cup R_i)$, such that $P \Rightarrow_{\mathcal{C}}^{+} Q$. If the critical pair is not contained in any set E_k, then by fairness it has to be contained in some set $PCP(E_j, R_j)$, which implies that there is a decomposition $P_1, \ldots, P_n$ of $u[P'\sigma]_p$. Since $P \succeq_{\mathcal{C}} u[P'\sigma]_p \succ_{\mathcal{C}} P_i$, for all i with $1 \leq i \leq n$, we may use the induction hypothesis to infer that each proof P_i can be transformed to a rewrite proof Q_i in R_∞. Let Q' be the composition $Q_1 \cdots Q_n$ of all these rewrite proofs. Since all terms in Q' are strictly smaller than u, we have $P \succ_{\mathcal{C}} Q$ and may apply the induction hypothesis again, to conclude that there is a rewrite proof of $s \approx t$ in R_∞, which in turn implies that there is some proof Q in $E_j \cup R_j$, such that $P \Rightarrow_{\mathcal{C}}^{+} Q$.

In summary, any proof P which is not a rewrite proof in R_∞ can be transformed to a simpler proof Q by $\Rightarrow_{\mathcal{C}}^{+}$. By the induction hypothesis, the proof Q, and hence P, can be transformed to a rewrite proof.   Q.E.D.

A special case of compositeness is the following criterion suggested by Kapur, Musser, and Narendran (1988).

Definition 2.38. Let R be a rewrite system. A critical overlap $t\sigma \leftarrow_{t \approx s}^{\lambda} s\sigma \rightarrow_{u \approx v}^{p} s\sigma[v\sigma]_p$ is said to be *prime* if no proper subterm of $s\sigma|_p = u\sigma$ is reducible by R.

A critical pair is prime if its corresponding critical overlap is prime.

For example, if R contains rewrite rules $(x^- \cdot y)^- \rightarrow y^- \cdot x^{--}$, $x \cdot x^- \rightarrow e$, and $x^{--} \rightarrow x$, then the first two rules define a critical overlap

$$ x^{---} \cdot x^{--} \leftarrow_R (x^- \cdot x^{--})^- \rightarrow_R e^-, $$

which is not prime, because the subterm x^{--} in $x^- \cdot x^{--}$ is reducible. Kapur, Musser, and Narendran (1988) proved the soundness of a criterion based on prime critical pairs, while Bachmair and Dershowitz (1988) established its correctness. In fact, all (correct) criteria mentioned above are special cases of the criterion PCP.

Various techniques can be used in practice to establish compositeness. Let P be a critical overlap $t\sigma \leftarrow_{t \approx s}^{\lambda} s\sigma \rightarrow_{u \approx v}^{p} s\sigma[v\sigma]_p$ and suppose the overlapped term $s\sigma = s\sigma[u\sigma]_p$ is reducible, say $s\sigma \rightarrow_{s' \approx t'}^{p'} w$.

If $p' = pq$ for some position q, and either $q \neq \lambda$ or else both $s \rhd s'$ and $u \rhd s'$, then there is a binary decomposition $P_1 P_2$ of P, where P_1 is the peak $t\sigma \xleftarrow{\lambda}_{t \approx s} s\sigma \xrightarrow{p'}_{s' \approx t'} w$ and P_2 the peak $w \xleftarrow{p'}_{t' \approx s'} s\sigma \xrightarrow{p}_{u \approx v} s\sigma[v\sigma]_p$. We have $P \succ_c P_1$ because

$$\{(\{s\sigma\}, s\sigma, s, t\sigma), (\{s\sigma\}, u\sigma, u, s\sigma[v\sigma])\}$$
$$\succ^c_{mul}$$
$$\{(\{s\sigma\}, s\sigma, s, t\sigma), (\{s\sigma\}, u\sigma|_q, s', w)\}$$

and $P \succ_c P_2$ because

$$\{(\{s\sigma\}, s\sigma, s, t\sigma), (\{s\sigma\}, u\sigma, u, s\sigma[v\sigma])\}$$
$$\succ^c_{mul}$$
$$\{(\{s\sigma\}, u\sigma|_q, s', w), (\{s\sigma\}, u\sigma, u, s\sigma[v\sigma])\}.$$

It should be noted that $s \rhd s'$ indicates that the rule $s \to t$ can be collapsed (and hence no critical pairs need to be computed with it). In practice it is therefore sufficient to require $q \neq \lambda$, which is the version of binary decomposition that was introduced by Kapur, Musser, and Narendran (1988). Note that such a decomposition exists whenever a critical pair is non-blocked.

If the position p' is not below q, additional information is required to ensure compositeness. Suppose that $p' \neq \lambda$ and there is a proof P_1 of $t\sigma \approx w$, such that $P \succ_c P_1$. Define P_2 to be the peak $w \leftarrow_R s\sigma \to_R s\sigma[v\sigma]$, to obtain a decomposition $P_1 P_2$ of the original peak. The difficulty, of course, is to determine whether a suitable proof P_1 exists. The techniques mentioned above for checking for connectedness may for instance be employed for that purpose.

A critical pair criterion may considerably decrease the total number of critical pairs generated by completion. This advantage may be offset, however, by the additional cost of checking whether the criterion applies to a given critical pair. Experimental results that give some indication of the utility of some critical pair criteria have been reported by Kapur, Musser and Narendran (1988) and by Küchlin (1985).

We conclude this section by establishing a Church-Rosser result via compositeness. We say that a critical pair $t\sigma \approx s\sigma[v\sigma]_p$ of a rule $u \to v$ on $s \to t$ at position p is *subsumed* by another critical pair $t\sigma' \approx s\sigma'[v'\sigma']_{p'}$ of a rule $u' \to v'$ on the same rule $s \to t$ at a position p' if (i) $p' \neq \lambda$ and (ii) there exists a substitution τ, such that $x\sigma = (x\sigma')\tau$, for all variables x occurring in s. We say that $S \subseteq CP(R)$ is a *complete set of critical pairs* if each critical pair in $CP(R)$ is subsumed by some critical pair in S.

Proposition 2.39. *Let R be a terminating rewrite system and S be a complete subset of $CP(R)$. If R is convergent on S, then R is Church-Rosser.*

Proof. It is sufficient to prove that all critical pairs in $CP(R) \setminus S$ are composite. Let $t\sigma \approx s\sigma[v\sigma]$ be a critical pair in $CP(R) \setminus S$, with corresponding overlap $P = t\sigma \leftarrow^{\lambda}_{t \approx s} s\sigma \rightarrow^{p}_{u \approx v} s\sigma[v\sigma]_p$. Since S is complete, the corresponding critical pair is subsumed by another critical pair $t\sigma' \approx s\sigma'[v'\sigma']_{p'}$ of some rule $u' \rightarrow v'$ on $s \rightarrow t$ at position p', where $p' \neq \lambda$ and there exists a substitution τ, such that $(t\sigma')\tau = t\sigma$ and $(s\sigma'[v'\sigma']_{p'})\tau = s\sigma[v\sigma]_{p'}$. Since R is convergent on S, there is a rewrite proof of $t\sigma' \approx s\sigma'[v'\sigma']_{p'}$. Let P_1 be a corresponding rewrite proof of $t\sigma \approx s\sigma[v\sigma]_{p'}$ and let P_2 be the proof $s\sigma[v'\sigma]_{p'} \leftarrow^{p'}_{u' \approx v'} s\sigma \rightarrow^{p}_{u \approx v} s\sigma[v\sigma]_p$. Evidently, $P \succ_c P_1$. Moreover, $P \succ_c P_2$ because

$$\{((\{s\sigma\}, s\sigma, s, t\sigma), (\{s\sigma\}, u\sigma, u, s\sigma[v\sigma])\}$$
$$\succ^c_{mul}$$
$$\{((\{s\sigma\}, u'\sigma, u', t\sigma), (\{s\sigma\}, u\sigma, u, s\sigma[v\sigma])\}$$

(note that $u'\sigma$ is a proper subterm of $s\sigma$). Therefore, $P_1 P_2$ is a decomposition of P.   Q.E.D.

This proposition has applications to rewrite systems containing rules $t[s, s] \rightarrow u$ with multiple occurrences of the same subterm on the left-hand side. For each critical pair obtained by superposing on one occurrence of s, there is a corresponding critical pair obtained by superposing on another occurrence of s. As all these critical pairs subsume each other, it is sufficient to compute only one of them. Such a critical pair criterion appears to be particularly useful in associative-commutative completion, where rewrite rules with multiple occurrences of identical subterms are common (Zhang and Kapur, 1989).

Summary

We have reformulated the Knuth-Bendix completion procedure as an equational inference system and have introduced proof orderings as a technique for reasoning about this and related equational inference systems. We have also formulated refinements of completion based on critical pair criteria in this framework. The key to this approach is the interpretation of completion as a proof normalization process. We have designed proof transformations to describe the effect of the inference rules on the proof level and defined a suitable measure of the complexity of proofs to show that proof transformations are normalizing. The notion of fairness characterizes those sequences of proof transformations for which normal-form proofs are rewrite proofs.

3. EXTENDED COMPLETION

Standard completion cannot be applied to equations, such as commutativity, for which the corresponding rewrite relation is non-terminating. There are also limitations in applications to theories with associativity or similar so-called permutative equations, as such theories often cannot be represented by *finite* convergent rewrite systems. For instance, from given rules

$$f(f(x,y),z) \rightarrow f(x,f(y,z))$$
$$f(a,b) \rightarrow b$$
$$f(a,f(x,b)) \rightarrow f(x,b)$$

completion may generate an infinite set of rules

$$f(a,f(x,f(y,b))) \rightarrow f(x,f(y,b))$$
$$f(a,f(x,f(y,f(z,b)))) \rightarrow f(x,f(y,f(z,b)))$$
$$\cdots$$

as pointed out by Peterson and Stickel (1981). Associativity and commutativity are typical equations that are more naturally viewed as "structural" axioms (defining a congruence relation on terms) rather than as "simplifiers" (defining a terminating rewrite relation).

In such situations it is often more fruitful to construct, not necessarily a convergent rewrite system, but instead a rewrite system for which normal forms are unique only up to the congruence defined by given structural axioms, such as associativity and commutativity. More precisely, given a set of equations E, a set of structural axioms A is singled out, and then a rewrite system R is constructed, such that $A \cup R$ defines the same equational theory as E, and normal forms with respect to R are unique up to the congruence generated by A. Huet (1980) proposed a corresponding completion procedure which applies to left-linear rewrite systems.

A more drastic approach relaxes not only the convergence requirement, but in addition extends the underlying rewrite relation to congruence classes of terms. Such extended completion procedures employ generalized versions of matching and unification that are defined with respect to the given congruence. Their advantage is that they also apply to non-left-linear systems.

For example, if rewriting takes into account the associativity axiom $f(f(x,y),z) \approx f(x,f(y,z))$, then the system consisting of the two rules $f(a,b) \to b$ and $f(a,f(x,b)) \to f(x,b)$ defines unique normal forms (up to associativity).

Extended completion procedures have been developed for various congruences. Basic concepts of the approach were described by Lankford and Ballantyne (1977a, b, c) and Peterson and Stickel (1981), for rewriting modulo associativity and commutativity. Jouannaud (1983) and Jouannaud and Kirchner (1986) demonstrated that a modified version of the Peterson-Stickel procedure can be applied not only to associativity and commutativity but to any equational theory with a finite, complete unification algorithm, provided all congruence classes are finite.

In this chapter we formalize these approaches within the inference system *cum* proof complexity framework. We describe the left-linear-rule method and present an extended completion method that covers arbitrary congruences, not just those for which congruence classes are finite. Before we discuss these methods, we introduce some basic terminology.

3.1. Rewriting Modulo a Congruence

Let A be a given set of equations. (For simplicity, we assume that A is symmetric, so that the two rewrite relations $\to_A$ and $\leftrightarrow_A$ are identical.) By $[s]$ we denote the congruence class of s with respect to A; that is, $[s]$ is the set $\{t : s \leftrightarrow_A^* t\}$. For any rewrite system R, we denote by R/A (R *mod* A) the set of all rewrite rules $s \to t$, such that $s \leftrightarrow_A^* u \to_R v \leftrightarrow_A^* t$, for some terms u and v. The system R/A represents the rewrite relation induced by R on congruence classes of A.

If congruence classes of A are finite (and an algorithm for enumerating the elements of each congruence class is given), then reducibility of terms by R/A is decidable, provided reducibility by R is decidable. To determine whether a term t is reducible by R/A, simply check whether some term in the congruence class of t is reducible by R. Since rewriting by R/A tends to be inefficient (if decidable at all), a slightly weaker, but often more practical notion of rewriting is usually employed within completion.

We say that a term t *A-matches* a term s if there is a substitution σ, such that $s\sigma \leftrightarrow_A^* t$. The system R_A consists of all rules $s \to t$, such that $s \leftrightarrow_A^* u\sigma$ and $t = v\sigma$, for some rule $u \to v$ in R and some substitution σ. The system R_A is contained in R/A. It generalizes rewriting by R in that it employs A-matching rather than ordinary matching. In this sense we speak of *rewriting modulo a congruence* or *extended rewriting*. A term t is reducible by R_A if and only if some subterm of t A-matches a left-hand side of a rule in R.

For example, if A consists of the associativity and commutativity axioms for addition, and R contains rules $-x + x \to 0$ and $f(x, x) \to x$, then $f(x + y, y + x)$ is irreducible by R, but can be reduced to $x + y$ by R_A. The term $-x + (x + y)$ is irreducible by R_A, but can be reduced to $0 + y$ by R/A.

A reduction ordering $\succ$ is said to be *compatible* with A if $u \leftrightarrow_A^* s \succ t \leftrightarrow_A^* v$ implies $u \succ v$, for all terms s, t, u, and v. Any ordering $\succ$ that is compatible with A induces an ordering (also denoted by $\succ$) on congruence classes of $\leftrightarrow_A^*$: $[s] \succ [t]$ if $s \succ t$. (This definition is valid precisely because $\succ$ is compatible with A.) A rewrite system R/A is terminating if and only if R is contained in some reduction ordering $\succ$ compatible with A.

A rewrite system R is said to be *Church-Rosser modulo A* if for all terms s and t with $s \leftrightarrow_{A \cup R}^* t$, there exist terms u and v, such that $s \to_R^* u \leftrightarrow_A^* v \leftarrow_R^* t$. We say that R is *convergent modulo A* (or *A-convergent*) if R/A is terminating and R is Church-Rosser modulo A. An A-convergent system defines normal forms that are unique up to the congruence $\leftrightarrow_A^*$.

If R is convergent modulo A and R' is any rewrite system with $R \subseteq R' \subseteq R/A$, then R' is also convergent modulo A and furthermore a term is irreducible by R' if and only if it is irreducible by R/A. In particular, if R_A is convergent modulo A, then it defines the same normal forms as R/A. Therefore R_A can be used instead of R/A for the purpose of normal-form computation.

The following example, given by Huet (1980), illustrates these notions. Let A be the set consisting of the associativity and commutativity axioms

$$
\begin{array}{rcl|rcl}
x + y & \approx & y + x & x \cdot y & \approx & y \cdot x \\
x + (y + z) & \approx & (x + y) + z & x \cdot (y \cdot z) & \approx & (x \cdot y) \cdot z
\end{array}
$$

and let R be the rewrite system

$$
\begin{array}{rcl|rcl}
x + 0 & \to & x & x \cdot 1 & \to & x \\
0 + x & \to & x & 1 \cdot x & \to & x \\
f(0) & \to & 1 & f(x + y) & \to & f(x) \cdot f(y)
\end{array}
$$

The system R/A is terminating. (Termination can be proved by an associative path ordering, Bachmair and Plaisted 1985.) Furthermore, R is Church-Rosser modulo A (a test for the Church-Rosser property is described below).

In the remaining sections of this chapter, we shall study methods for constructing a rewrite system R for a given set of equations E, such that $A \cup E$ and $A \cup R$ define the same equational theory and a certain rewrite system R', with $R \subseteq R' \subseteq R/A$, is convergent modulo A. (Two cases will be considered: $R' = R$ and $R' = R_A$.)

3.2.   The Left-Linear Rule Method

Following the approach outlined in the previous chapter, we discuss Huet's method for constructing A-convergent rewrite systems from the perspective of proof normalization. We assume that all equations in A are of the form $s \approx_0 t$, while other equations are labelled by the symbol $\top$ or $\bot$. As before, equations $s \approx_\bot t$ are called rewrite rules and written $s \to t$.

In this context, a normal-form proof is a *rewrite proof modulo A in R*; that is, a proof of the form $s \to_R^* v \leftrightarrow_A^* w \leftarrow_R^* t$. As before, by a peak we mean a proof $s \leftarrow_R u \to_R t$. By a *cliff* we mean a proof of the form $s \leftrightarrow_A u \to_R t$ or $s \leftarrow_R u \leftrightarrow_A t$.

A rewrite system R is Church-Rosser modulo A if and only if there is a rewrite proof modulo A in R for every equation provable in $A \cup R$. A proof in $A \cup R$, on the other hand, is a rewrite proof modulo A if and only if it contains neither a peak nor a cliff. The Critical Pair Lemma indicates that a peak $s \leftarrow_R u \to_R t$ can either be replaced either by a rewrite proof $s \to_R^* v \leftarrow_R^* t$ or by a critical-pair proof $s \leftrightarrow_{CP(R)} t$.

Let P be a cliff $s \leftrightarrow_A u \to_R t$. If P is a proper overlap, then by the Critical Pair Lemma we have $s \leftrightarrow_{CP(A \cup R) \setminus CP(A)} t$. If P is a non-overlap, then a rewrite proof modulo A can be obtained by commuting the two proof steps. That is, there is a term v, such that $s \to_R v \leftrightarrow_A t$. Also, any variable overlap $s \leftrightarrow_A u \to_R t$ in which the second proof step applies below the first, can be replaced by a rewrite proof modulo A of the form $s \to_R^* v \leftrightarrow_A w \leftarrow_R^* t$ (cf. the Critical Pair Lemma). Unfortunately, similar variable overlaps in which the first proof step applies below the second can be problematic, as there may exist only a proof $s \leftrightarrow_A^* v \to_R w \leftrightarrow_A^* t$, but not necessarily a rewrite proof modulo A.

For example, the equation $a \approx b$ and the rule $x \cdot x \to x$ determine a variable overlap

$$a \cdot b \leftrightarrow_A a \cdot a \to_R a$$

that can only be replaced by a proof

$$a \cdot b \leftrightarrow_A b \cdot b \to_R b \leftrightarrow_A a$$

but not by a rewrite proof modulo A. The first two proof steps

$$a \cdot b \leftrightarrow_A b \cdot b \to_R b$$

in this proof form a variable overlap that can be replaced by a proof

$$a \cdot b \leftrightarrow_A a \cdot a \to_R a \leftrightarrow_A b$$

which contains the initial variable overlap as a subproof.

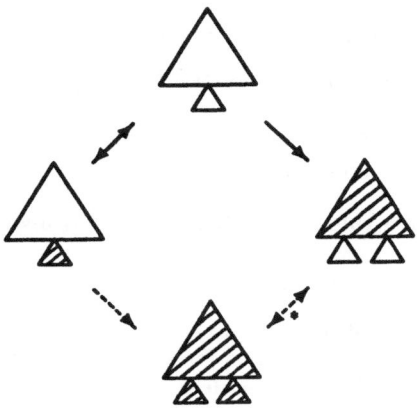

Figure 3.1: *Left-linear variable overlap*

This example involves a *non-left-linear* rewrite rule. In fact, a variable overlap $s \leftrightarrow_A u \rightarrow_R t$, wherein the first proof step is below the second, can always be replaced by a rewrite proof modulo A, if the proof step $u \rightarrow_R t$ is by application of a *left-linear* rewrite rule (see Figure 3.1 and compare with Figure 2.3). In short, the proof simplification techniques we have outlined in the previous chapter can be directly applied to extended completion, provided all (persisting) rewrite rules are left-linear.

Let $\succ$ be a reduction ordering that is compatible with A. The inference system $\mathcal{L}_A^\succ$ differs from the standard completion system $\mathcal{C}^\succ$ only in the following inference rules:

DEDUCTION:
$$\frac{E \,;\, R}{E \cup \{s \approx t\} \,;\, R} \qquad \text{if } s \approx t \in CP(A \cup R)$$

DELETION:
$$\frac{E \cup \{s \approx s\} \,;\, R}{E \,;\, R} \qquad \text{if } s \leftrightarrow_A^* t$$

where R may be any rewrite system ordered by $\succ$. The inference rules for orientation, simplification, composition, and collapse are the same as for standard completion. Thus, if $A = \emptyset$, then $\mathcal{L}_A^\succ$ is essentially the same as the standard completion system $\mathcal{C}^\succ$. (The only difference is that the present deduction rule, being formulated in terms of critical pairs, is more restrictive than the corresponding inference rule of standard completion.)

Inference systems $\mathcal{L}_A$ are called *left-linear completion systems*. They are sound in the following sense.

Lemma 3.1. *Whenever* $E \, ; R \vdash_{\mathcal{L}_A} E' \, ; R'$, *then the two congruence relations* $\leftrightarrow_{A \cup E \cup R}$ *and* $\leftrightarrow_{A \cup E' \cup R'}$ *are the same.*

The above deduction rule is reflected by proof transformation rules

$$s \leftarrow_R u \rightarrow_R t \;\; \Rightarrow \;\; s \leftrightarrow_E t$$
$$s \leftrightarrow_A u \rightarrow_R t \;\; \Rightarrow \;\; s \leftrightarrow_E t$$

while the new deletion rule is reflected by proof transformation rules

$$s \leftrightarrow_E t \;\; \Rightarrow \;\; s \leftrightarrow_A^* t.$$

The remaining transformation rules, for orientation, simplification, composition, and collapse, are the same as for standard completion.

By $\mathcal{R}_{\mathcal{L}}^{\succ}$ (or simply $\mathcal{R}_{\mathcal{L}}$, if $\succ$ is clear from the context) we denote the set of all these transformation rules. The corresponding transformation relation is denoted by $\Rightarrow_{\mathcal{L}}$.

Lemma 3.2. *The proof transformation system* $\mathcal{R}_{\mathcal{L}}$ *reflects the inference system* $\mathcal{L}$.

Lemma 3.3. *The transformation relation* $\Rightarrow_{\mathcal{L}}$ *induced by* $\mathcal{R}_{\mathcal{L}}$ *is terminating.*

Proof. We define a proof ordering $\succ_{\mathcal{L}}$ to prove termination of $\Rightarrow_{\mathcal{L}}$. The ordering is based on suitable a measure of the complexity of a proof. The cost of a proof step $s \leftrightarrow_{u \approx_n v}^p t$ is defined to be:

$$\begin{cases} (\{[s], [t]\}, \top, \bot) & \text{if } n = \top \\ (\{[s], [t]\}, \bot, \bot) & \text{if } n = 0 \\ (\{[s]\}, u, \{[t]\}) & \text{if } n = \bot \text{ and } u \succ v \\ (\{[t]\}, v, \{[s]\}) & \text{if } n = \bot \text{ and } v \succ u \end{cases}$$

The complexity of a proof is the multiset of all costs of its proof steps. Again, a proof P has the same complexity as the inverse proof P^{-1}.

Let $\succ^l$ be the lexicographic combination of the multiset extension of the reduction ordering $\succ$, the encompassment ordering $\rhd$, and the reduction ordering $\succ$. We assume that the symbols $\bot$ and $\top$ are minimum and maximum elements in any ordering. Proofs are compared with respect to their complexity, using the multiset extension of the ordering $\succ^l$. The resulting ordering, which is a proof ordering, is denoted by $\succ_{\mathcal{L}}$. It contains the transformation system $\mathcal{R}_{\mathcal{L}}$ (and hence $\Rightarrow_{\mathcal{L}}$). The proof is similar to the proof of Lemma 2.20 in the case of those transformation rules that have been directly taken from standard completion—orientation, simplification, composition, and collapse. We discuss the remaining cases in detail.

i) *Deduction.* We have $s \leftarrow_R u \rightarrow_R t \succ_{\mathcal{L}} s \leftrightarrow_E t$, because $u \succ s$ and $u \succ t$. Furthermore, if $u \succ t$ then

$$(\{[s],[u]\}, \perp, \perp) \succ^{l} (\{[s],[t]\}, \top, \perp),$$

which implies $s \leftrightarrow_A u \rightarrow_R t \succ_{\mathcal{L}} s \leftrightarrow_E t$.

ii) *Deletion.* We have $s \leftrightarrow_E t \succ_{\mathcal{L}} s \leftrightarrow_A^* t$, because $(\{[s],[t]\}, \top, \perp) \succ^{l}$ $(\{[u],[v]\}, \perp, \perp$, whenever $s \leftrightarrow_A^* u$ and $v \leftrightarrow_A^* t$. Q.E.D.

In analogy to Lemma 2.11 (and using similar arguments in the proof) we obtain:

Lemma 3.4. *If P is a peak $s \leftarrow_R u \rightarrow_R t$, a cliff $s \leftrightarrow_A u \rightarrow_R t$, or a cliff $s \leftarrow_R u \leftrightarrow_A t$, and Q is a rewrite proof modulo A of $s \approx t$, then $P \Rightarrow_{\mathcal{L}}^* Q$.*

The transformation relation $\Rightarrow_{\mathcal{L}}$ also provides a way to prove the following Church-Rosser theorem.

Theorem 3.5. (Huet 1980) *Let R be a rewrite system and A be a set of equations, such that R/A terminates. The system R is Church-Rosser modulo A if and only if, for all terms s and t with $s \leftarrow_R u \rightarrow_{A \cup R} t$, there exist terms v and w, such that $s \rightarrow_R^* v \leftrightarrow_A^* w \leftarrow_R^* t$.*

Proof. If R is Church-Rosser modulo A, then evidently any peak or cliff can be replaced by a rewrite proof modulo A.

On the other hand, suppose that every peak or cliff can be replaced by a rewrite proof modulo A. Let $\succ$ be the reduction ordering $\rightarrow_{R/A}^+$ induced by R/A and $\Rightarrow_{\mathcal{L}}^+$ be the transformation relation associated with $\mathcal{R}_{\mathcal{L}}^{\succ}$. We prove by induction on $\Rightarrow_{\mathcal{L}}^+$ that every proof in $A \cup R$ can be transformed into a rewrite proof modulo A in R. If a proof P in $A \cup R$ is not a rewrite proof modulo A, then it must contain a peak $s \leftarrow_R u \rightarrow_R t$, a cliff $s \leftrightarrow_A u \rightarrow_R t$, or a cliff $s \leftarrow_R u \leftrightarrow_A t$. By our assumption, for any such peak or cliff there exists a rewrite proof modulo A of the form $s \rightarrow_R^* v \leftrightarrow_A^* w \leftarrow_R^* t$. Using Lemma 3.4, we may infer that there is a proof Q, such that $P \Rightarrow_{\mathcal{L}}^+ Q$. By the induction hypothesis, the proof Q (and hence P) can be transformed into a rewrite proof modulo A. Q.E.D.

Let us emphasize that this theorem applies to arbitrary rewrite systems R, not just left-linear systems. Left-linearity will be crucial, however, in deriving effective conditions under which all cliffs can be transformed to rewrite proofs modulo A.

By $\mathcal{N}_{\mathcal{L}}^{\succ}$ we denote the set of all rewrite proofs modulo A. We can now formulate a sufficient condition for fairness.

Lemma 3.6. *A non-failing derivation* $E_0 ; R_0 \vdash_{\mathcal{L}} E_1 ; R_1 \vdash_{\mathcal{L}} \cdots$ *is fair with respect to* $\mathcal{R}_{\mathcal{L}}$ *and* $\mathcal{N}_{\mathcal{L}}$ *if* R_∞ *is left-linear and the set of critical pairs* $CP(A \cup R_\infty) \setminus CP(A)$ *is a subset of the set of all deduced equations* $\bigcup_k E_k$.

Proof. Let $E_0 ; R_0 \vdash_{\mathcal{L}} E_1 ; R_1 \vdash_{\mathcal{L}} \cdots$ be a derivation for which $E_\infty = \emptyset$, R_∞ is left-linear, and $CP(A \cup R_\infty) \setminus CP(A)$ is a subset of $\bigcup_k E_k$. We have to show that whenever a proof P in $A \cup E_\infty \cup R_\infty$ contains a peak $s \leftarrow_{R_\infty} u \rightarrow_{R_\infty} t$, a cliff $s \leftrightarrow_A u \rightarrow_{R_\infty} t$, or a cliff $s \leftarrow_{R_\infty} u \leftrightarrow_A t$, then there is a proof Q in $\bigcup_i A \cup E_i \cup R_i$, such that $P \Rightarrow_{\mathcal{L}}^+ Q$.

By the Critical Pair Lemma, any peak or cliff in P which is a non-overlap or variable overlap can be replaced by a rewrite proof modulo A. (As we have pointed out above, it is essential in the variable-overlap case that all rules in R_∞ be left-linear.) Since $CP(A \cup R_\infty) \setminus CP(A)$ is a subset of $\bigcup_k E_k$, every proper overlap can be replaced by a simpler proof, from which the assertion can be inferred easily.   Q.E.D.

Lemmas 3.2, 3.3, and 3.6 indicate that Theorem 2.10 can be applied to the inference system $\mathcal{L}$, the proof transformation system $\mathcal{R}_{\mathcal{L}}$, and the set of proofs $\mathcal{N}_{\mathcal{L}}$.

Theorem 3.7. *Let* R_0 *be a rewrite system and* $\succ$ *be a reduction ordering that contains* R_0 *and is compatible with* A. *If* $E_0 ; R_0 \vdash_{\mathcal{L}} E_1 ; R_1 \vdash_{\mathcal{L}} \cdots$ *is a derivation for which* $E_\infty = \emptyset$, R_∞ *is left-linear, and* $CP(A \cup R_\infty) \setminus CP(A)$ *is a subset of* $\bigcup_k E_k$, *then* R_∞ *is convergent modulo* A.

The main limitation of the above completion method is that only left-linear convergent rewrite systems can be produced. Rewrite systems with rules such as $x^- \cdot x \rightarrow 0$ or $x \cdot x \rightarrow x$ are not within its scope. Indeed, to eliminate variable overlaps that result from non-left-linear rules, so-called "variable instance pairs" would have to be computed (see Bachmair, Dershowitz, and Hsiang 1986).

For instance, the variable overlap $a \cdot b \leftrightarrow_A a \cdot a \rightarrow_R a$, determined by the equation $a \approx b$ and the rule $x \cdot x \rightarrow x$, defines a variable instance pair $a \cdot b \approx a$. More precisely, the equation $a \cdot b \approx a$ is obtained by superposing the equation $a \approx b$ on the *instance* $a \cdot a \rightarrow a$ of the rule $x \cdot x \rightarrow x$.

Exhaustive computation of variable instance pairs is impractical in general. An alternative, which we shall pursue in the remaining sections of this chapter, is to approximate rewriting on congruence classes by the extended rewrite relation R_A. Then each problematic variable overlap can be regarded as a single rewrite step. For instance, the variable overlap $a \cdot b \leftrightarrow_A a \cdot a \rightarrow_R a$ corresponds to an extended rewrite step $a \cdot b \rightarrow_{R_A} a$. Rewriting modulo a congruence thus avoids problematic variable overlaps, but presents the difficulty of eliminating more general overlaps involving the extended rewrite relation R_A. For many sets of equations A, such extended overlaps can be effectively eliminated.

3.3. Church-Rosser Systems

We first consider under which conditions an extended rewrite system R_A is convergent modulo A. By $\mathcal{N}_{\mathcal{E}}$ we denote the set of all rewrite proofs modulo A in R_A, that is, proofs of the form $s \to^*_{R_A} u \leftrightarrow^*_A v \leftarrow^*_{R_A} t$, where R is a rewrite system contained in the given reduction ordering $\succ$. (The reduction ordering is assumed to be compatible with A, so that R/A is terminating.)

A rewrite system R_A is Church-Rosser modulo A if and only if there is a rewrite proof modulo A in R_A for every equation valid in $A \cup R_A$. A proof in $A \cup R_A$, on the other hand, is a rewrite proof modulo A in R_A if and only if it contains no peak $s \leftarrow_{R_A} u \to_{R_A} t$ and no cliff $s \leftrightarrow_A u \to_{R_A} t$ (or the inverse of such a cliff). If R_A is Church-Rosser modulo A, then every peak or cliff can be replaced by a rewrite proof modulo A. If R/A is terminating, the converse direction is also true. That is, if R_A is not Church-Rosser modulo A, then some peak or cliff can not be replaced by a rewrite proof modulo A.

The Critical Pair Lemma, when applied to the rewrite system R_A, indicates that non-overlaps and variable overlaps can always be replaced by rewrite proofs, while proper overlaps $s \leftarrow_{R_A} u \to_{R_A} t$ can be replaced by proofs $s \leftrightarrow_{CP(R_A)} t$. Since the set $CP(R_A)$ is in general infinite, this does not yield a practical method for dealing with proper overlaps. (Of course, the same difficulty also arises with cliffs $s \leftrightarrow_A u \to_{R_A} t$.) We shall therefore deal with systems R_A indirectly, via R, by regarding $s \to_{R_A} t$ not as an elementary proof step in R_A, but instead as a proof $s \leftrightarrow^*_A s' \to_R t$ in $A \cup R$.

More precisely, by $s \leftrightarrow^{\leq p}_E t$ we denote a proof

$$s = u_0 \leftrightarrow^{p_1}_{e_1} u_1 \leftrightarrow^{p_2}_{e_2} u_2 \leftrightarrow^{p_3}_{e_3} \cdots u_{n-1} \leftrightarrow^{p_n}_{e_n} u_n = t$$

where all positions $p_1, \ldots, p_n$ are below p and all equations $e_1, \ldots, e_n$ are in E. Similarly, we write $s \leftrightarrow^{<p}_E t$, if the positions $p_1, \ldots, p_n$ are strictly below p. By $s \to_{R_A} t$ we mean a proof $s \leftrightarrow^{\leq p}_A u \to^p_e t$, where e is an equation in R. Thus, when speaking of a peak $s \leftarrow_{R_A} u \to_{R_A} t$, we are referring to a proof $s \leftarrow^p_e s' \leftrightarrow^{\leq p}_A u \leftrightarrow^{\leq q}_A t' \to^q_{e'} t$, where e and e' are equations in R.

Consider such a peak

$$s \leftarrow^p_e s' \leftrightarrow^{\leq p}_A u \leftrightarrow^{\leq q}_A t' \to^q_{e'} t,$$

where the position p is not strictly below q. This proof can be written either as

$$s \leftarrow^p_e s' \leftrightarrow^{\leq q}_A t' \to^q_{e'} t$$

or as

$$s \leftarrow^p_e s' \leftrightarrow^{\leq p}_A u \leftrightarrow^{p'}_{e''} u' \leftrightarrow^{\leq q}_A t' \rightarrow^q_{e'} t,$$

where e'' is an equation in A and the position p' is not below q.

Henceforth, by a peak $s \leftarrow_R u \rightarrow_{R_A} t$ we intend a proof

$$s \leftarrow^p_e u \leftrightarrow^{\leq q}_A t' \rightarrow^q_{e'} t,$$

where p is not strictly below q; by a cliff $s \leftrightarrow_A u \rightarrow_{R_A} t$, a proof

$$s \leftrightarrow^p_e u \leftrightarrow^{\leq q}_A t' \rightarrow^q_{e'} t$$

where p is not below q.

The above considerations indicate that a proof is a rewrite proof modulo A in R_A if and only if it contains no peak or cliff (or the inverse thereof). We shall next describe which peaks and cliffs can always be replaced by rewrite proofs modulo A and describe techniques to eliminate "problematic" peaks and cliffs.

A peak $s \leftarrow^p_{w \approx v} u \leftrightarrow^{\leq q}_A t' \rightarrow^q_{v' \approx w'} t$, where p is not strictly below q, is called a *non-overlap* if p and q are disjoint positions. On the other hand, if $q = pp'$, for some position p', then P is called a (extended) *proper overlap* if p' is a non-variable position in v; and a (extended) *variable overlap*, otherwise. The same classification is applied to cliffs $s \leftrightarrow^p_e u \leftrightarrow^{\leq q}_A t' \rightarrow^q_{e'} t$ where p is not below q. If such a cliff is a variable overlap or proper overlap, then q is a position strictly below p. Non-overlaps and variable overlaps can be replaced by rewrite proofs modulo A in the following way.

A non-overlap $s \leftarrow^p_{w \approx v} u \leftrightarrow^{\leq q}_A t' \rightarrow^q_{v' \approx w'} t$ is a proof of the form $u[Q]_p u[Q']_q$, where Q is the proof $s|_p \leftarrow^\lambda_{w \approx v} u|_p$ and Q' the proof $u|_q \leftrightarrow^*_A t'|_q \rightarrow^\lambda_{v' \approx w'} t|_q$. The alternative proof $s[Q']_q t[Q]_p$ is a rewrite proof modulo A. We associate the transformation rule

$$u[Q]_p u[Q']_q \quad \Rightarrow \quad s[Q']_q t[Q]_p$$

with the non-overlap.

If $s \leftarrow^p_{w \approx v} u \leftrightarrow^{\leq q}_A t' \rightarrow^q_{v' \approx w'} t$ is a variable overlap, then $q = pq'q''$, where $v|_{q'} = x$, for some variable x. Let σ be a substitution, such that $u|_p = v\sigma$ and $s|_p = w\sigma$. Let Q be the subproof $s \leftarrow^p_{w \approx v} u$ and Q' be the proof

$$x\sigma = u|_{pq'} \leftrightarrow^{\leq q''}_A t'|_{pq'} \rightarrow^{q''}_{v' \approx w'} t|_{pq'} = u'$$

so that the variable overlap is the composition of Q with the proof $u[Q']_{pq'}$.

Let now $p_1, \ldots, p_n$ be all the positions in w at which the variable x occurs; and let $s_0 = s$ and $s_i = s[u', \ldots, u']_{pp_1, \ldots, pp_i}$, for all i with $1 \leq i \leq n$. By P_i we denote the proof $s_{i-1}[Q']_{pp_i}$. Note that P_i is a proof of $s_{i-1} \approx s_i$.

Similarly, let $q_1, \ldots, q_m$ be all the positions in v, other than q', at which x occurs; and let $t_0 = t$ and $t_i = t[u', \ldots, u']_{pq_1, \ldots, pq_i}$, for all i with $1 \leq i \leq m$. By Q_i we denote the proof $t_{i-1}[Q']_{pq_i}$. Note that Q_i a proof of $t_{i-1} \approx t_i$.

Furthermore, let ρ be a substitution for which $x\rho = u'$ and $y\rho = y\sigma$, if $y \neq x$. Then $t_m|_p = v\rho$ and $s_n = t_m[w\rho]_p$. Consequently, there is a proof $Q'' = s_n \leftrightarrow_{w \approx v}^p t_m$.

We associate with the variable overlap the proof transformation

$$ Qu[Q']_{pq'} \quad \Rightarrow \quad P_1 \cdots P_n Q'' Q_m \cdots Q_1 $$

which transforms the variable overlap into a rewrite proof modulo A. (Note that n is the number of occurrences of x in w and $m+1$ is the number of occurrences of x in v.)

The above proof transformations apply to non-overlaps and variable overlaps—peaks and cliffs. If complete sets of A-unifiers can be computed, then proper overlaps can also be dealt with effectively.

Two terms s and t are said to be A-unifiable if there exists a substitution σ, called an A-unifier, such that $s\sigma \leftrightarrow_A^* t\sigma$. A set Σ of A-unifiers of s and t is said to be complete if for every A-unifier τ of s and t there exist substitutions σ in Σ and ρ, such that $x\tau \leftrightarrow_A^* (x\sigma)\rho$, for all variables x occurring in s or t. (We may assume without loss of generality that whenever σ is an element of a complete set of A-unifiers of s and t, then $x\sigma = x$, for all variables x not occurring in s or t.) A set Σ of A-unifiers of s and t is said to be minimal if, given any two distinct substitutions σ and τ in Σ, there is no substitution ρ, such that $x\tau \leftrightarrow_A^* (x\sigma)\rho$, for all variables x in s or t.

Finite, complete sets of unifiers need not exist for arbitrary sets of equations (Plotkin 1972). There are also theories for which minimal, complete sets of unifiers do not always exist (Fages and Huet 1986). However, if there is a finite, complete set of unifiers, then there is a minimal one. Minimal, complete sets of unifiers are also unique in a certain sense (for details see Fages and Huet 1986). Algorithms for computing minimal, complete sets of unifiers are known for many theories of practical importance, including commutativity (Plotkin 1972); associativity and commutativity (Stickel 1981; Fages 1987); and associativity, commutativity, and identity (Fages 1987). If A is the empty set, then the set consisting of a most general unifier of s and t is complete. For a survey on unification, see Siekmann (1984). More recent results and further references can be found in Kirchner (1986, 1988) and Gallier and Snyder (1990).

Henceforth, we shall assume that A is a set of equations for which minimal complete sets of unifiers exist.

Definition 3.8. Let $u \approx v$ and $s \approx t$ be equations with no variables in common (if necessary, the variables of one equation are renamed) and let p be a non-variable position in u, such that $u|_p$ and s are A-unifiable with minimal complete set of unifiers Σ.[1] By an *A-critical overlap* we mean any proof $v\sigma \leftarrow^\lambda_{v\approx u} u\sigma[s\sigma]_p \leftrightarrow^{\leq p}_A u\sigma[s\sigma]_p \rightarrow^p_{s\approx t} u\sigma[t\sigma]_p$, where $\sigma \in \Sigma$. The equation $v\sigma \approx u\sigma[t\sigma]_p$ is called an *A-critical pair*.

In other words, A-critical pairs differ from ordinary critical pairs in that A-unification is used instead of unification. By $CP_A(E)$ we denote the set of all A-critical pairs between equations in E. By $CP_A(E, A)$ we denote the set of all A-critical pairs of equations in E on equations in A.

By the *(left) variable part* of a proof step $s \leftrightarrow^p_{s'\approx t'} t$ we mean the set of all positions pqr in s, for which $s'|_q$ is a variable. (The *right variable part* is the set of all positions pqr in t, for which $t'|_q$ is a variable.)

Lemma 3.9. (Extended Critical Pair Lemma, Jouannaud 1983) *Let R be a set of equations and suppose complete sets of A-unifiers can be computed.*

(1) If $s \leftarrow_{A\cup R} u \rightarrow_{R_A} t$ is a non-overlap or variable overlap, then there exists a rewrite proof $s \rightarrow^*_{R_A} v \leftrightarrow^*_A w \leftarrow^*_{R_A} t$.

(2) If $s \leftarrow_{A\cup R} u \rightarrow_{R_A} t$ is a proper overlap, then there exists a proof $s \leftrightarrow^*_A s' \leftrightarrow_{CP_A(R)\cup CP_A(R,A)} t' \leftrightarrow^*_A t$ in which all proof steps in $s \leftrightarrow^*_A v$ apply in the variable part of $s \leftarrow_{A\cup R} u$.

Proof. We have already proved the first part of the lemma. For the second part, let P be a proper overlap $s \leftarrow^p_{w\approx v} u \leftrightarrow^{\leq q}_A t' \rightarrow^q_{v'\approx w'} t$, where $q = pq'$, for some position q'. (We may assume, without loss of generality, that $v \approx w$ and $v' \approx w'$ have no variables in common.) Let σ be a substitution, such that $u|_p = v\sigma$ and $u|_q = v'\sigma$. Since $v|_{q'}$ is A-unifiable with v', there is a complete set of A-unifiers Σ. Let $\tau \in \Sigma$ and ρ be substitutions, such that $w\tau \approx v\tau[w'\tau]_{q'}$ is an A-critical pair and $x\sigma \leftrightarrow^*_A x\tau\rho$, for all variables x in $v \approx w$ and $v' \approx w'$. As a consequence, we have $s|_p \leftrightarrow^*_A w\tau\rho$ and $t|_p \leftrightarrow^*_A v\tau\rho[w'\tau\rho]_{q'}$, which implies that there is a suitable proof $s \leftrightarrow^*_A s' \leftrightarrow_{CP_A(R)\cup CP_A(R,A)} t' \leftrightarrow^*_A t$.　　　　Q.E.D.

The Extended Critical Pair Lemma serves as the basis for a Church-Rosser result for extended rewrite relations R_A.

Definition 3.10. The *subterm ordering modulo A* is defined by: $s \succeq_A t$ if some subterm of s is equivalent to t in A. The *encompassment ordering modulo A* is defined by: $s \rhd_A t$ if either $s \rhd t$ or else $s \leftrightarrow^{\leq\lambda}_A s[t\sigma]$, for some substitution σ.

---

[1] We may assume that for each σ in Σ and variable x in u or s, the term $x\sigma$ contains none of the variables occurring in $s \approx t$ or $u \approx v$.

In other words, $s \rhd_A t$ if either s is a proper instance of t or else some proper subterm of s is equivalent with respect to A to an instance of t. The relation $\rhd_A$ is not well-founded in general, but we have:

Lemma 3.11. *The encompassment ordering modulo A is well-founded if and only if the (strict part of the) subterm ordering modulo A is well-founded.*

Proof. Clearly, if the encompassment ordering modulo A is well-founded, so is the subterm ordering modulo A. On the other hand, suppose the encompassment ordering modulo A is not well-founded. Since the standard encompassment ordering is well-founded, there exists an infinite sequence of ground terms $t_1, t_2, t_3, \ldots$ and a corresponding sequence of positions $p_1, p_2, p_3, \ldots$ such that, for all i, $p_i \neq \lambda$ and $t_{i+1} \leftrightarrow_A^* t_i|_{p_i}$. Thus the subterm ordering modulo A is not well-founded.   Q.E.D.

Note that if the subterm ordering modulo A is well-founded, then all congruence classes of A have to be finite.

Theorem 3.12. (Jouannaud and Kirchner 1986) *Let A be a set of equations for which minimal complete sets of A-unifiers can be computed, and let R be a rewrite system such that R/A is terminating. If the subterm ordering modulo A is well-founded, then R_A is Church-Rosser modulo A if and only if for all A-critical pairs $s \approx t$ in $CP_A(R) \cup CP_A(R, A)$, there exist terms s' and t', such that $s \rightarrow_{R_A}^* s' \leftrightarrow_A^* t' \leftarrow_{R_A}^* t$.*

Proof. The only-if direction is trivial. To prove the if-direction we first define a well-founded ordering $\succ_{A,R}$ on proofs in $A \cup R$.

Let $\succ$ be the reduction ordering $\rightarrow_{R/A}^+$ induced by R/A and let P be a proof

$$t_0 \leftrightarrow_{e_1}^{p_1} t_1 \leftrightarrow_{e_2}^{p_2} t_2 \leftrightarrow_{e_3}^{p_3} \cdots t_{n-1} \leftrightarrow_{e_n}^{p_n} t_n$$

in $A \cup R$, where e_i is an equation $u_i \approx v_i$.

The i-th proof step in P is said to be *bound on the left* if P contains a subproof $t_{j-1} \leftarrow_{R_A} t_i$, and *bound on the right* if it contains a subproof $t_{i-1} \rightarrow_{R_A} t_j$. We define: $\beta_i^P = 0$, if the i-th proof step in P is bound on the left and right; $\beta_i^P = 1$, if it is either bound on the left or on the right, but not both; and $\beta_i^P = 2$, otherwise.

The complexity of single proof steps in P is defined by:

$$\begin{cases} (\{[t_{i-1}]\}, [t_{i-1}|_{p_i}], \beta_i^P) & \text{if } u_i \approx v_i \in A \\ (\{[t_{i-1}]\}, \bot, \bot) & \text{if } u_i \approx v_i \in R \\ (\{[t_i]\}, \bot, \bot) & \text{if } v_i \approx u_i \in R \end{cases}$$

The complexity of a proof is the collection of all tuples associated with its proof steps. Tuples are compared lexicographically, using the multiset extension $\succ_{mul}$ of the reduction ordering $\succ$ in the first component, the encompassment ordering modulo A in the second component, and the usual greater-than relation in the last component. Proofs are compared according to their complexity, using the multiset extension of this ordering on tuples. We denote the corresponding ordering on proofs by $\succ_{A,R}$. If the proper subterm ordering modulo A is well-founded, then $\succ_{A,R}$ is well-founded. In general, $\succ_{A,R}$ is not a proof ordering, as the third component of the complexity of a proof step depends on the context in which a proof step appears.

We show that all minimal (with respect to $\succ_{A,R}$) proofs in $A \cup R$ are rewrite proofs modulo A. If P is a minimal proof, but not a rewrite proof modulo A, it contains either a peak $s \leftarrow_R u \rightarrow_{R_A} t$ or a cliff $s \leftrightarrow_A u \rightarrow_{R_A} t$ (or the inverse of such a cliff).

Peaks. By the Extended Critical Pair Lemma, any peak $P' = s \leftarrow_R u \rightarrow_{R_A} t$ can be replaced by a proof P'' in which all terms are strictly smaller than u, so that $P' \succ_{A,R} P''$ and, as can easily be seen, also $P[P'] \succ_{A,R} P[P'']$.

Cliffs, non-overlaps. A non-overlap is a proof $P' = u[Q]_p u[Q']_q$, where Q is a proof $s|_p \leftarrow^\lambda_{w \approx v} u|_p$ in A and Q' is a proof $u|_q \leftrightarrow^*_A t'|_q \rightarrow^\lambda_{v' \approx w'} t|_q$ in $A \cup R$. Any such non-overlap can be replaced by a proof $P'' = s[Q']_q t[Q]_p$. Since $u[Q']_q$ has the same complexity as $s[Q']_q$ and $u[Q]_p \succ_{A,R} t[Q]_p$, we have $P' \succ_{A,R} P''$. It can easily be checked that also $P[P'] \succ_{A,R} P[P'']$.

Cliffs, variable overlaps. A variable overlap is a proof of the form $P' = Qu[Q']_{pq'}$ which, as we have shown above, can be replaced by a proof P'' of the form $P_1 \cdots P_n Q'' Q_m \cdots Q_1$. Since R/A is terminating, we have $n > 0$. Since $Q \succ_{A,R} P''$ we have $P' \succ_{A,R} P''$ and also $P[P'] \succ_{A,R} P[P'']$.

Cliffs, proper overlaps. By the Extended Critical Pair Lemma, every proper overlap P' of the form

$$s \leftarrow^p_{w \approx v} u \leftrightarrow^{\leq q}_A t' \rightarrow^q_{v' \approx w'} t,$$

where $v \approx w$ is an equation in A and $v' \approx w'$ is a rule in R, can be replaced by a "critical-pair proof" $s \leftrightarrow^*_A s' \leftrightarrow_{CP_A(R,A)} t' \leftrightarrow^*_A t$, where all proof steps in $s \leftrightarrow^*_A s'$ apply in the variable part of $s \leftarrow^p_{w \approx v} u$. Moreover, the stated assumptions guarantee that there is a rewrite proof modulo A of $s' \approx t'$.

Observe that w cannot be a variable, for otherwise the subterm ordering modulo A would not be well-founded. As a consequence, the variable part of $s \leftarrow^p_{w \approx v} u$ contains only positions strictly below p. Hence, there is a proof P'' of the form

$$s \leftrightarrow^{<p}_A s' \rightarrow^*_{R_A} s'' \leftrightarrow^*_A t'' \leftarrow^*_{R_A} t' \leftrightarrow^*_A t.$$

We have

$$s \leftarrow^p_{w \approx v} u \succ_{A,R} s \leftrightarrow^{\leq p}_A s'$$

(using the second proof complexity component);

$$s \leftarrow^p_{w \approx v} u \succ_{A,R} s' \rightarrow^*_{R_A} s''$$

(using all three proof complexity components); and

$$s \leftarrow^p_{w \approx v} u \succ_{A,R} s'' \leftrightarrow^*_A t'' \leftarrow^*_{R_A} t' \leftrightarrow^*_A t$$

(using the first proof complexity component). In sum, we have $P' \succ_{A,R} P''$ and, as can easily be checked, also $P[P'] \succ_{A,R} P[P'']$. Q.E.D.

Computation of A-critical pairs is the basis of extended completion procedures. While A-critical pairs are invariably used to eliminate peaks, different mechanisms have been suggested for elimination of cliffs. Consider a proper overlap

$$s \leftrightarrow^p_{w \approx v} u \leftrightarrow^{\leq q}_A t' \rightarrow^q_{v' \approx w'} t,$$

where $v \approx w$ is an equation in A and $v' \approx w'$ is a rewrite rule in R. Then $q = pq'$, for some position $q' \neq \lambda$, such that $v|_{q'}$ is not a variable and is A-unifiable with v'. Placing the rule $v' \approx w'$ in a larger context, we obtain a new rule $v[v']_{q'} \approx v[w']_{q'}$ and a corresponding proof

$$s \leftrightarrow^p_{w \approx v} u \leftrightarrow^{\leq q}_A t' \rightarrow^p_{v[v']_{q'} \approx v[w']_{q'}} t,$$

so that $s \rightarrow_{R'_A} t$, where $R' = R \cup \{v[v']_{q'} \approx v[w']_{q'}\}$. In other words, the cliff has been eliminated.

Definition 3.13. (Jouannaud and Kirchner 1986) Let $s \approx t$ be a rewrite rule and $u \approx v$ be an equation in A that has no variables in common with $s \approx t$, such that some proper non-variable subterm $u|_p$ is A-unifiable with s. Then $u[s]_p \approx u[t]_p$ is called an *extended rule* of $s \approx t$ with respect to A.

By $EXT_A(R)$ we denote the set of all extended rules of R with respect to A. Extended rules were originally introduced by Peterson and Stickel (1981) in the context of associative-commutative rewriting. We emphasize that computation of extended rules does not involve A-unification. It suffices to know that $u|_p$ and s are A-unifiable, but the actual A-unifiers need not be computed. Extended rules are therefore often easier to compute than A-critical pairs.

For example, a rewrite rule can be extended with respect to the associativity axiom $x + (y + z) \approx (x + y) + z$ if and only if its left-hand side is unifiable under associativity with $y + z$ or $x + y$, which is the case if the rule is of the form $s + t \rightarrow u$. The corresponding extended rules are $x + (s+t) \rightarrow x+u$ and $(s+t)+x \rightarrow u+x$. Note that there are no extended rules with respect to commutativity axioms $x+y \approx y+x$, as such equations contain no proper non-variable subterms.

3.4. Extended Completion

We will next formulate extended completion as an equational inference system. Some of the inference rules are designed so that equations can be kept fully simplified and redundant equations can be deleted. For instance, in the presence of $s \rightarrow t$ a rewrite rule $u[s] \rightarrow u[t]$ is usually redundant and can be removed. However, extended rules serve a special purpose and have to be exempted from this deletion mechanism. Therefore we partition a given rewrite system R into two sets N and S, where rules in S are protected from certain simplification mechanisms. Formally, an *unprotected rule* is an equation $s \approx_\perp t$, while a *protected rule* is an equation $s \approx_1 t$. Equations in A are of the form $s \approx_0 t$; all remaining equations are of the form $s \approx_\top t$. An inference rule, in this context, is a binary relation on tuples $E; N; S$, where N contains all unprotected rules and S all protected rules.

Let $\succ$ be a reduction ordering that is compatible with A. The inference system $\mathcal{E}_A^\succ$ (or simply $\mathcal{E}$) contains the following inference rules:

DEDUCTION: $$\frac{E; N; S}{E \cup \{s \approx t\}; N; S} \qquad \text{if } s \approx t \in CP_A(A \cup R)$$

EXTENSION: $$\frac{E; N; S}{E; N; S \cup \{s \rightarrow t\}} \qquad \text{if } s \approx t \in EXT_A(R)$$

ORIENTATION: $$\frac{E \cup \{s \simeq t\}; N; S}{E; N \cup \{s \rightarrow t\}; S} \qquad \text{if } s \succ t$$

PROTECTION: $$\frac{E; N \cup \{s \rightarrow t\}; S}{E; N; S \cup \{s \rightarrow t\}}$$

DELETION: $$\frac{E \cup \{s \approx t\}; N; S}{E; N; S} \qquad \text{if } s \leftrightarrow_A^* t$$

where $R = N \cup S$ denotes a rewrite system contained in $\succ$. The following inference rules, which are also part of $\mathcal{E}_A^\succ$, are indispensable for efficiency:

SIMPLIFICATION: $$\frac{E \cup \{s \simeq t\}; N; S}{E \cup \{u \simeq t\}; N; S} \qquad \text{if } s \rightarrow_{R/A} u$$

COMPOSITION:
$$\frac{E; N \cup \{s \to t\}; S}{E; N \cup \{s \to u\}; S} \qquad \text{if } t \to_{R/A} u$$

$$\frac{E; N; S \cup \{s \to t\}}{E; N; S \cup \{s \to u\}} \qquad \text{if } t \to_{R/A} u$$

COLLAPSE:
$$\frac{E; N \cup \{s \to t\}; S}{E \cup \{u \approx t\}; N; S} \qquad \begin{array}{l} \text{if } s \leftrightarrow_{A}^{\leq p} s' \to_{v \approx w}^{p} u, \\ \text{for some rule } v \to w \text{ in} \\ R, \text{where } s \gg v \end{array}$$

where $\gg$ may be any well-founded ordering on terms, e.g., the encompassment ordering $\rhd$. The difference between protected and unprotected rules is that collapse inferences may be applied only to the latter.

Inference systems $\mathcal{E}_A$ are called *extended completion systems*. Standard completion can be simulated by extended completion: let A be the empty set and $\gg$ be the encompassment ordering, and consider only tuples $E; N; \emptyset$.

Inference systems $\mathcal{E}$ are sound:

Lemma 3.14. *If $E; N; S \vdash_{\mathcal{E}_A} E'; N'; S'$, then the two congruence relations $\leftrightarrow_{A \cup E \cup N \cup S}^{*}$ and $\leftrightarrow_{A \cup E' \cup N' \cup S'}^{*}$ are the same.*

If $E; N; S \vdash_{\mathcal{E}} E'; N'; S'$ and the reduction ordering $\succ$ contains $N \cup S$, then $\succ$ also contains $N' \cup S'$ (recall that the ordering $\succ$ is compatible with A). Consequently, the limit rewrite system $(N_\infty \cup S_\infty)/A$ is terminating for any derivation for which the initial rewrite system $N_0 \cup S_0$ is contained in the reduction ordering $\succ$. Furthermore, if $E; N; S \vdash_{\mathcal{E}} E'; N'; S'$, then any term that is reducible by $(N \cup S)/A$ is also reducible by $(N' \cup S')/A$.

Definition 3.15. By an *A-completion procedure* we mean a program that accepts as input a reduction ordering $\succ$, a set of equations E_0, and a rewrite system $R_0 = N_0 \cup S_0$ contained in $\succ$; and uses the inference rules of $\mathcal{E}_A^{\succ}$ to generate a derivation from $E_0; N_0; S_0$.

Definition 3.16. We say that an *A-completion procedure fails* for a given input, if $E_\infty \neq \emptyset$. It is *correct*, if $(N_\infty \cup S_\infty)_A$ is convergent modulo A, whenever $E_\infty = \emptyset$.

We shall use proof orderings to derive sufficient conditions for the correctness of *A*-completion procedures.

3.5. The Extended Rule Method

Peterson and Stickel (1981) designed an *A*-completion procedure, for sets A of associativity and commutativity axioms, that is characterized by the systematic use of extended rules. We demonstrate that similar techniques can be applied to arbitrary sets of equations A for which minimal complete sets of unifiers exist. That is, we will show that suitable requirements regarding the computation of extended rules ensure the fairness and correctness of A-completion procedures.

First we design a suitable proof transformation system reflecting extended completion. Let A be the given set of equations, $\succ$ be a reduction ordering compatible with A, and $\gg$ be a well-founded ordering.

Extension is reflected by transformation rules

$$u[s]_p \to^p_{s\approx_\perp t} u[t]_p \quad \Rightarrow \quad u[s]_p \to^\lambda_{u[s]\approx_1 u[t]} u[t]_p$$
$$u[s]_p \to^p_{s\approx_1 t} u[t]_p \quad \Rightarrow \quad u[s]_p \to^\lambda_{u[s]\approx_1 u[t]} u[t]_p$$

where $s \succ t$ and $p \neq \lambda$.

Orientation and protection are reflected by transformation rules

$$s \to^\lambda_{s\approx_\top t} t \quad \Rightarrow \quad s \to^\lambda_{s\approx_\perp t} t$$
$$s \to^\lambda_{s\approx_\perp t} t \quad \Rightarrow \quad s \to^\lambda_{s\approx_1 t} t$$

where $s \succ t$.

Deletion is reflected by

$$s \leftrightarrow_E t \quad \Rightarrow \quad s \leftrightarrow^*_A t.$$

Simplification is reflected by

$$s \leftrightarrow_E t \quad \Rightarrow \quad s \to_{R/A} u \leftrightarrow_E t$$

(where $s \to_{R/A} u$ denotes a proof $s \leftrightarrow^*_A s' \to_R u' \leftrightarrow^*_A u$).

Composition is reflected by

$$s \to^\lambda_{s\approx_\perp t} t \quad \Rightarrow \quad s \to^\lambda_{s\approx_\perp u} u \leftarrow_{R/A} t$$
$$s \to^\lambda_{s\approx_1 t} t \quad \Rightarrow \quad s \to^\lambda_{s\approx_1 u} u \leftarrow_{R/A} t$$

where $s \succ t$ and $s \succ u$.

Collapse is reflected by

$$s \to^\lambda_{s\approx_\perp t} t \quad \Rightarrow \quad s \leftrightarrow^{\leq p}_A s' \to^p_{v\approx_n w} u \leftrightarrow^\lambda_{u\approx_\top t} t$$

where $s \succ t$, $v \succ w$, $s \gg v$, and either $n = \perp$ or $n = 1$.

The computation of A-critical pairs of $CP_A(R)$, which is part of deduction, is reflected by transformation rules

$$s \leftarrow_R u \to_{R_A} t \quad \Rightarrow \quad s \leftrightarrow^*_A v \leftrightarrow_E w \leftrightarrow^*_A t.$$

The other part of deduction—the computation of A-critical pairs of $CP_A(R, A)$—is not required with the extended rule method, as all cliffs that are proper overlaps are eliminated by extended rules.

By $\mathcal{R}_{\mathcal{E}}^{\succ}$ we denote the set of all these proof transformation rules, plus the transformation rules for elimination of cliffs that are non-overlaps or variable overlaps. (The transformation rules for elimination of cliffs have to be included, even though they do not reflect any inference rules.) By $\Rightarrow_{\mathcal{E}}$ we denote the corresponding transformation relation.

Lemma 3.17. *If P is a peak $s \leftarrow_R u \rightarrow_{R_A} t$ and Q is a rewrite proof modulo A of the form $s \rightarrow^*_{R/A} v \leftrightarrow^*_A w \leftarrow^*_{R/A} t$, then $P \Rightarrow^+_{\mathcal{E}} Q$.*

The analogous property for cliffs $s \leftrightarrow_A u \rightarrow_{R_A} t$ does not hold, as the transformation relation $\Rightarrow_{\mathcal{E}}$ is too weak. It is an open question whether $\mathcal{R}_{\mathcal{E}}$ can be extended to a terminating transformation system in which a cliff can always be transformed to a corresponding rewrite proof modulo A.

Lemma 3.18. *The transformation system $\mathcal{R}_{\mathcal{E}}$ reflects the extended completion system $\mathcal{E}$.*

Lemma 3.19. *The transformation system $\mathcal{R}_{\mathcal{E}}$ is terminating.*

Proof. We design a well-founded ordering $\succ_{\mathcal{E}}$ that contains $\Rightarrow_{\mathcal{E}}$. Let P be a proof
$$t_0 \leftrightarrow^{p_1}_{e_1} t_1 \leftrightarrow^{p_2}_{e_2} t_2 \leftrightarrow^{p_3}_{e_3} \cdots t_{n-1} \leftrightarrow^{p_n}_{e_n} t_n$$
where e_i is an equation $u_i \approx_{n_i} v_i$.

The i-th proof step in P is said to be *bound on the left* if, for some $j < i$, we have $n_j = 1$, while $n_{j+1} = \cdots n_i = 0$ and all positions $p_{j+1}, \ldots, p_i$ are below p_j. (In other words, P contains a subproof $t_{j-1} \leftarrow_{S_A} t_i$.) The proof step is *bound on the right* if, for some $j > i$, we have $n_i = \cdots n_{j-1} = 0$, $n_j = 1$, and all positions $p_i, \ldots, p_{j-1}$ are below p_j. (That is, P contains a subproof $t_{i-1} \rightarrow_{S_A} t_i$.) We define: $\beta^P_i = 0$, if the i-th proof step in P is bound on the left and right; $\beta^P_i = 1$, if it is either bound on the left or on the right, but not both; and $\beta^P_i = 2$, otherwise.

The complexity of single proof steps in P is defined by:

$$
\begin{cases}
(\{[t_{i-1}], [t_i]\}, \bot, \bot, \bot, \bot) & \text{if } n_i = \top \\
(\{[t_{i-1}]\}, u_i, \bot, \bot, [t_i]) & \text{if } n_i = \bot \text{ and } u_i \succ v_i \\
(\{[t_i]\}, v_i, \bot, \bot, [t_{i-1}]) & \text{if } n_i = \bot \text{ and } v_i \succ u_i \\
(\{[t_{i-1}]\}, \bot, \beta^P_i, \bot, [t_i]) & \text{if } n_i = 0 \\
(\{[t_{i-1}]\}, \bot, \bot, |p_i|, [t_i]) & \text{if } n_i = 1 \text{ and } u_i \succ v_i \\
(\{[t_i]\}, \bot, \bot, |p_i|, [t_{i-1}]) & \text{if } n_i = 1 \text{ and } v_i \succ u_i
\end{cases}
$$

where $|p|$ denotes the length of the sequence p (i.e., the depth of the redex in the term) and β^P_i is as defined previously. The complexity of a proof is the collection of all tuples associated with its proof steps.

Complexity tuples are compared lexicographically by the multiset extension $\succ_{mul}$ of the reduction ordering $\succ$ (in the first component), the ordering $\rhd$ (in the second component), the greater-than relation on the natural numbers (in the third and fourth component), and the reduction ordering $\succ$ (in the last component). We denote this lexicographic ordering by $\succ^e$. Proofs are compared according to their complexity, using the multiset extension of the ordering $\succ^e$. The corresponding ordering, which we denote by $\succ_{\mathcal{E}}$, is well-founded. Moreover, if $P \succ_{\mathcal{E}} Q$, then $u[P\sigma]_p \succ_{\mathcal{E}} u[Q\sigma]_p$, for all terms u, positions in u, and substitutions σ. However, since the complexity of a proof step depends on the context in which it appears, $Q \succ_{\mathcal{E}} Q'$ does not necessarily imply $P[Q] \succ_{\mathcal{E}} P[Q']$. Therefore $\succ_{\mathcal{E}}$ is not a proof ordering.

We will first show that all transformation rules of $\mathcal{R}_{\mathcal{E}}$ are decreasing with respect to the above complexity measure. The transformation rules reflecting orientation, simplification, and elimination of peaks decrease the complexity of a proof in the first component. The transformations rules reflecting collapse reduce the complexity in the second component; the transformations rules reflecting extension reduce it in the second or fourth component. Protection and composition reduce the complexity in the fourth and fifth component, respectively.

i) *Extension.* We have $u[s]_p \to^p_{s\approx_\perp t} u[t]_p \succ_{\mathcal{E}} u[s]_p \to^\lambda_{u[s]\approx_1 u[t]} u[t]_p$ because $(\{[u[s]_p]\}, s, \perp, \perp, [u[t]_p]) \succ^e (\{[u[s]_p]\}, \perp, \perp, |p|, [u[t]_p])$.

Similarly, if $p \neq \lambda$, then $u[s]_p \to^p_{s\approx_1 t} u[t]_p \succ_{\mathcal{E}} u[s]_p \to^\lambda_{u[s]\approx_1 u[t]} u[t]_p$ because $(\{[u[s]_p]\}, \perp, \perp, |p|, [u[t]_p]) \succ^e (\{[u[s]_p]\}, \perp, \perp, 0, [u[t]_p])$.

ii) *Orientation.* We have $s \to^\lambda_{s\approx_\top t} t \succ_{\mathcal{E}} s \to^\lambda_{s\approx_\perp t} t$ because $\{[s], [t]\} \succ_{mul} \{[s]\}$.

iii) *Protection.* By similar arguments as in the first case of extension.

vi) *Deletion.* We have $s \leftrightarrow_E t \succ_{\mathcal{E}} s \leftrightarrow^*_A t$ because $\{[s], [t]\} \succ_{mul} \{[u]\}$, for each term u in $s \leftrightarrow^*_A t$.

v) *Simplification.* We have $s \leftrightarrow_E t \succ_{\mathcal{E}} s \to_{R/A} u \leftrightarrow_E t$ because $\{[s], [t]\} \succ_{mul} \{[v]\}$, for each term v in $s \to_{R/A} u$, and $\{[s], [t]\} \succ_{mul} \{[u], [t]\}$.

vi) *Composition.* We have $s \to^\lambda_{s\approx_\perp t} t \succ_{\mathcal{E}} s \to^\lambda_{s\approx_\perp u} u \leftarrow_{R/A} t$ because $(\{[s]\}, s, \perp, \perp, [t]) \succ^e (\{[s]\}, s, \perp, \perp, [u])$ and in addition s is bigger than each term in $u \leftarrow_{R/A} t$.

Also, $s \to^\lambda_{s\approx_1 t} t \succ_{\mathcal{E}} s \to^\lambda_{s\approx_1 u} u \leftarrow_{R/A} t$ because $(\{[s]\}, \perp, \perp, 0, [t]) \succ^e (\{[s]\}, \perp, \perp, 0, [u])$ and s is bigger than each term in $u \leftarrow_{R/A} t$.

vii) *Collapse.* If $s \rhd v$, then $s \to^\lambda_{s\approx_\perp t} t \succ_{\mathcal{E}} s \leftrightarrow^{\leq p}_A s' \to^p_{v\approx_n w} u \leftrightarrow^\lambda_{u\approx_\top t} t$ because the rewrite step $s \to^\lambda_{s\approx_\perp t} t$ is more complex than all proof steps in $s \leftrightarrow^*_A s' \to^p_{v\approx_n w} u$ (in the second component) and also more complex than $u \leftrightarrow^\lambda_{u\approx_\top t} t$ (in the first component).

viii) *Peaks.* We have $s \leftarrow_R u \to_{R_A} t \succ_{\mathcal{E}} s \leftrightarrow^*_A v \leftrightarrow_E w \leftrightarrow^*_A t$ because u

is bigger than each term in $s \leftrightarrow_A^* v \leftrightarrow_E w \leftrightarrow_A^* t$.

viii) *Cliffs, non-overlaps.* By the same arguments as in the proof of Theorem 3.12.

ix) *Cliffs, variable overlaps.* Let P be a variable overlap $s \leftarrow_{w \approx_0 v}^p u \leftrightarrow_A^{\leq q} t' \rightarrow_{v' \approx_n w'}^q t$, where $q = pq'q''$ and $v|_{q'} = x$, for some variable x, and either $n = \perp$ or $n = 1$. In addition, $v' \succ w'$. Let Q be the proof $s \leftarrow_{w \approx_0 v}^p u$ and Q' the proof

$$u|_{pq'} \leftrightarrow_A^{\leq q''} t'|_{pq'} \rightarrow_{v' \approx_n w'}^{q''} t|_{pq'}$$

so that $P = Qu[Q']_{pq'}$. Then P can be replaced by a rewrite proof modulo A of the form $P' = P_1 \cdots P_k Q'' Q_m \cdots Q_1$, where all proofs P_i and Q_j are of the form $u' \rightarrow_{R_A} u''$ (cf. the proof of Theorem 3.12).

Since the rewrite system R/A is terminating, we have $k > 0$. Then all terms in the proof $P_2 \cdots P_k Q'' Q_m \cdots Q_1$ are strictly smaller than s, so that $Q \succ_\mathcal{E} P_2 \cdots P_k Q'' Q_m \cdots Q_1$. In addition, if $n = \perp$ (i.e., the rewrite step in Q' is by application of an unprotected rule), then $u[Q']_{pq'}$ and P_1 have the same complexity. On the other hand, if $n = 1$ (i.e., the rewrite step in Q' is by application of a protected rule), then $u[Q']_{pq'}$ is the composition of two proofs $Q^{(1)} = u \leftrightarrow_A^{\leq q} t'$ and $Q^{(2)} = t' \rightarrow_{v' \approx_1 w'}^q t$. Similarly, the proof P_1 is the composition of two proofs $P_1^{(1)} = s \leftrightarrow_A^{\leq q} s_1'$ and $P_1^{(2)} = s_1' \rightarrow_{v' \approx_1 w'}^q s_1$. Then $P_1^{(1)}$ has the same complexity as $Q^{(1)}$ and $Q \succ_\mathcal{E} P_1^{(2)}$. (The third component of the complexity ensures that proof steps by protected rules are simpler than proof steps in A. It would be sufficient to use a constant instead of β_i^P, but the slightly more complicated complexity measure will be of advantage in the case of sets A of associativity and commutativity axioms discussed below.) In either case, we have $P \succ_\mathcal{E} P'$.

We have now proved that all transformation rules $Q \Rightarrow Q'$ of $\mathcal{R}_\mathcal{E}$ are decreasing with respect to the given complexity measure. Thus we also have $u[Q\sigma]_p \succ_\mathcal{E} u[Q'\sigma]_p$, for all terms u, positions in u, and substitutions σ. It can be shown that in all the cases under consideration, we also have $P[Q] \succ_\mathcal{E} P[Q']$, which implies that $\mathcal{R}_\mathcal{E}$ is terminating. Q.E.D.

We can now characterize (non-failing) derivations that induce normalizing sequences of proof transformations.

Lemma 3.20. *A non-failing derivation* $E_0; N_0; S_0 \vdash_\mathcal{E} E_1; N_1; S_1 \vdash_\mathcal{E} \cdots$ *is fair with respect to* $\mathcal{R}_\mathcal{E}$ *and* $\mathcal{N}_\mathcal{E}$ *if the set of critical pairs* $CP_A(R_\infty)$ *is a subset of* $\bigcup_k E_k$ *and the set of extended rules* $EXT_A(R_\infty)$ *is a subset of* $\bigcup_k S_k$.

Proof. Let $E_0; N_0; S_0 \vdash_\mathcal{E} E_1; N_1; S_1 \vdash_\mathcal{E} \cdots$ be a derivation for which $E_\infty = \emptyset$, $CP_A(R_\infty)$ is a subset of $\bigcup_k E_k$, and $EXT_A(R_\infty)$ is a subset of $\bigcup_k S_k$.

We have to show that whenever a proof P in $A \cup E_\infty \cup R_\infty$ contains a peak or cliff, then there is a proof Q in $A \cup \bigcup_i (E_i \cup R_i)$, such that $P \Rightarrow_{\mathcal{E}}^+ Q$.

Since $E_\infty = \emptyset$, let us assume P is a proof in $A \cup R_\infty$, but contains a peak $s \leftarrow_{R_\infty} u \rightarrow_{(R_\infty)_A} t$ or a cliff $s \leftarrow_A u \rightarrow_{(R_\infty)_A} t$. Any peak or cliff which is a non-overlap or a variable overlap can be transformed to a rewrite proof modulo A by $\Rightarrow_{\mathcal{E}}^+$.

If P contains a proper overlap $s \leftarrow_{R_\infty} u \rightarrow_{(R_\infty)_A} t$, then by the Extended Critical Pair Lemma $s \leftrightarrow_A^* s' \leftrightarrow_{CP_A(R_\infty)} t' \leftrightarrow_A^* t$. Since by fairness $CP_A(R_\infty) \subseteq \bigcup_k E_k$, there is a proof $s \leftrightarrow_A^* s' \leftrightarrow_{E_k} t' \leftrightarrow_A^* t$, for some $k \geq 0$. Consequently, there is a suitable proof Q such that $P \Rightarrow_{\mathcal{E}}^+ Q$.

Finally, suppose P contains a proper overlap $s \leftrightarrow_A u \rightarrow_{(R_\infty)_A} t$. By fairness, $EXT_A(R_\infty)$ is a subset of $\bigcup_j S_j$, which implies that there is a proof $s \rightarrow_{(S_k)_A} t$, for some k. Again, we may infer that $P \Rightarrow_{\mathcal{E}}^+ Q$, for some proof Q in $A \cup \bigcup_i (E_i \cup R_i)$.   Q.E.D.

A completion procedure is said to be *fair with respect to extensions* if it generates only derivations for which $CP_A(R_\infty)$ is a subset of $\bigcup_k E_k$ and $EXT_A(R_\infty)$ is a subset of $\bigcup_k S_k$. Applying Theorem 2.10 to the inference system $\mathcal{E}$, the rewrite system $\mathcal{R}_{\mathcal{E}}$, and the set of proofs $\mathcal{N}_{\mathcal{E}}$, we obtain:

Theorem 3.21. *Let A be a set of equations with a minimal complete unification algorithm. If an A-completion procedure is fair with respect to extensions and does not fail for the given input, then $(N_\infty \cup S_\infty)_A$ is convergent modulo A.*

Fairness with respect to extensions guarantees the correctness of extended completion and in this sense characterizes a class of correct completion procedures. Fair procedures are required (1) to compute all A-critical pairs between persisting rules and (2) to extend and *protect* all persisting rules. Protection of persisting rules may necessitate the computation of infinitely many extended rules. The requirement can be weakened, though, for many theories A that are of interest in practice, as we shall demonstrate for the case of associativity and commutativity.

3.6. Associative-Commutative Completion

By an associative-commutative completion procedure we mean any AC-completion procedure, where AC is a set of associativity and commutativity axioms

$$f(x, f(y, z)) \approx f(f(x, y), z)$$
$$f(f(x, y), z) \approx f(x, f(y, z))$$
$$f(x, y) \approx f(y, x)$$

for some operator symbols f (called AC operators). Moreover, we assume that the ordering $\gg$ (used in the definition of the collapse inference rule) is defined by: $s \gg t$ if s is equivalent in AC to some term containing an instance of t, but not vice versa. (This ordering is indeed well-founded.)

The only extended rules with respect to associativity and commutativity originate from rules $f(s,t) \rightarrow u$ with an AC operator f as outermost symbol on the left-hand side. They are $f(x, f(s,t)) \rightarrow f(x, u)$ and $f(f(s,t), x) \rightarrow f(u, x)$, where x is a new variable not appearing in s, t, or u.

Unfortunately, extended rules themselves can be extended. For instance, the extended rule $f(x, f(s,t)) \rightarrow f(x, u)$ spans two further extended rules $f(y, f(x, f(s,t))) \rightarrow f(y, f(x, u))$ and $f(f(x, f(s,t)), y) \rightarrow f(f(x, u), y)$. Indeed, since extended rules cannot be collapsed, no derivation producing a finite system R_∞ that contains a rule $f(s,t) \rightarrow u$ can be fair with respect to extensions. (That is, if $R_\infty = N_\infty \cup S_\infty$ contains a single rule with left-hand side $f(s,t)$, then it contains rules with arbitrarily large left-hand sides.) Fortunately, we shall be able to establish the correctness of AC-completion procedures under considerably weaker requirements.

Definition 3.22. Let R be a rewrite system. We define R^e as the rewrite system consisting of R plus all extensions $f(s,x) \rightarrow f(t,x)$ of rules $s \rightarrow t$ in R for which there exist no rule $u \rightarrow v$ in R and substitution σ, such that $f(s,x) \leftrightarrow^*_{AC} u\sigma$ and $f(t,x) \rightarrow^*_{AC \cup R/AC} v\sigma$.

The set R^e is evidently a subset of $R \cup EXT_{AC}(R)$. It describes the extended rules that are actually computed in the Peterson-Stickel procedure.

For example, let R be the set consisting of three rules, $x + 0 \rightarrow x$, $x + -x \rightarrow 0$ and $x * 0 \rightarrow 0$. The set R^e consists of R plus a single extended rule $(x + -x) + y \rightarrow 0 + y$. The extended rule $(x + 0) + y \rightarrow x + y$ is not in R^e, because it is equivalent under AC to the instance $(x + y) + 0 \rightarrow x + y$ of $x + 0 \rightarrow x$. Similarly, since $(x * 0) * y \leftrightarrow^*_{AC} (x * y) * 0$ and $0 * y \rightarrow_{R/AC} 0$, the extended rule $(x * 0) * y \rightarrow 0 * y$ is not in R^e either.

Lemma 3.23. $(R^e)^e = R^e$, for every rewrite system R.

Proof. Let $s \rightarrow t$ be a rewrite rule in R^e. We show that there exist a rule $u \rightarrow v$ in R^e and a substitution σ, such that $f(s,x) \leftrightarrow^*_{AC} u\sigma$ and $f(t,x) \rightarrow^*_{AC \cup R/AC} v\sigma$.

If $s \rightarrow t$ is a rule in R, then the assertion follows from the definition of R^e. If $s \rightarrow t$ is not in R, it is an extended rule $f(s', y) \rightarrow f(t', y)$. Let σ be a substitution, such that $y\sigma = f(y, x)$ and $z\sigma = z$, for all variables z different from y. Then $s'\sigma = s'$ and $t'\sigma = t'$, so that

$$f(s, x) = f(f(s', y), x) \leftrightarrow^*_{AC} f(s'\sigma, y\sigma)$$

and

$$f(t, x) = f(f(t', y), x) \leftrightarrow^*_{AC} f(t'\sigma, y\sigma).$$

In other words, every extension of an extended rule is equivalent in AC to an instance of the extended rule itself. Q.E.D.

The essential property of sets R^e is expressed in the following lemma (cf. the notion of "AC-compatibility", Peterson and Stickel 1981).

Lemma 3.24. *Let $f(s, t) \to u$ be a rule in R for which f is an AC operator. For every term v and substitution σ, there exist terms v' and w and a rule $s' \approx t'$ in R^e, such that $f(f(s\sigma, t\sigma), v) \leftrightarrow^*_{AC} v' \to^\lambda_{s' \approx t'} w$ and $f(u\sigma, v) \to^*_{AC\cup R/AC} w$.*

Proof. Consider the extended rule $f(f(s, t), x) \to f(u, x)$. By the definition of R^e, there exist a rule $s' \to t'$ in R^e and a substitution τ, such that $f(f(s, t), x) \leftrightarrow^*_{AC} s'\tau$ and $f(u, x) \to^*_{ACUR/AC} t'\tau$. We may assume without loss of generality that $x\sigma$ is v. Thus, we have $f(f(s\sigma, t\sigma), v) \leftrightarrow^*_{AC} (s'\tau)\sigma$ and $f(u\sigma, v) \to^*_{ACUR/AC} (t'\tau)\sigma$. Since $s' \to t'$ is in R^e, the assertion follows immediately. Q.E.D.

Proposition 3.25. *Let AC be a set of associativity-commutativity axioms and $E_0; N_0; S_0 \vdash_{\mathcal{E}_{AC}} E_1; N_1; S_1 \vdash_{\mathcal{E}_{AC}} \cdots$ be a derivation, such that (i) $E_\infty = \emptyset$, (ii) $CP_{AC}(R_\infty)$ is a subset of $\bigcup_k E_k$, and (iii) R^e_∞ is a subset of $\bigcup_k S_k$. Then $(R_\infty)_{AC}$ is convergent modulo AC.*

Proof. Let $E_0; N_0; S_0 \vdash_{\mathcal{E}_{AC}} E_1; N_1; S_1 \vdash_{\mathcal{E}_{AC}} \cdots$ be a derivation satisfying the stated conditions. It suffices to show that whenever a proof P in $AC \cup R_\infty$ contains a peak or a cliff (or the inverse thereof), then there is a proof Q in $AC \cup \bigcup(E_j \cup R_j)$, such that $P \succ_{\mathcal{E}} Q$.

By the same arguments as in Lemma 3.20, it can be shown that whenever P contains either a peak or else a cliff which is a non-overlap or a variable overlap, then there is a suitable proof Q, such that $P \succ_{\mathcal{E}} Q$. Suppose P contains a proper overlap $s \leftrightarrow_{AC} u \to_{(R_\infty)_{AC}} t$. We may assume without loss of generality that this cliff P' is of the form

$$\begin{aligned} f(f(u, v), w) \quad &\leftrightarrow_{AC} \quad f(u, f(v, w)) \\ &\leftrightarrow^*_{AC} \quad f(u, f(s\sigma, t\sigma)) \quad \to_{f(s,t) \approx_n u'} \quad f(u, u'\sigma) \end{aligned}$$

or

$$\begin{aligned} f(v, f(w, u)) \quad &\leftrightarrow_{AC} \quad f(f(v, w), u) \\ &\leftrightarrow^*_{AC} \quad f(f(s\sigma, t\sigma), u) \quad \to_{f(s,t) \approx_n u'} \quad f(u'\sigma, u), \end{aligned}$$

where $f(s,t) \approx_n u'$ is a rule in R_∞.

By Lemma 3.24, there exist terms v' and w' and a rule $s' \approx t'$ in R_∞^e, such that $f(f(s\sigma, t\sigma), u) \leftrightarrow^*_{AC} v' \to^\lambda_{s' \approx t'} w'$ and $f(u'\sigma, u) \to^*_{AC \cup R_\infty^e / AC} w'$. In other words, both $f(f(u,v),w)$ and $f(v, f(w,u))$ are reducible by $(R_\infty^e)_{AC}$ at the top.

Let us consider one of the two cases, the other is similar. By fairness, R_∞^e is a subset of $\bigcup_k S_k$, so that the proof

$$P'' = f(f(u,v),w) \to_{(R_\infty)_{AC}} w' \leftarrow^*_{AC \cup (R_\infty)/AC} f(u, u'\sigma)$$

is a proof in $AC \cup \bigcup_k R_k$. We have

$$f(f(u,v),w) \leftrightarrow_{AC} f(u, f(v,w)) \succ_\varepsilon f(f(u,v),w) \to_{(R_\infty)_{AC}} w'$$

(all AC steps in the latter proof are bound on the right) and

$$f(f(u,v),w) \leftrightarrow_{AC} f(u, f(v,w)) \succ_\varepsilon w' \leftarrow^*_{AC \cup (R_\infty)/AC} f(u, u'\sigma)$$

(all terms in the latter proof are strictly smaller than $f(u, f(v,w))$). In sum, there is a suitable proof Q, such that $P \succ_\varepsilon Q$.   Q.E.D.

In contrast to Theorem 3.21, the above proposition requires R_∞^e, instead of $EXT(R_\infty)$, to be a subset of $\bigcup S_k$. The essential point, of course, is that R_∞^e is finite, whenever R_∞ is finite.

Let us also point out that AC-critical pairs obtained by superposing on a proper subterm of an extended rule are superfluous. Namely, if there is a proper overlap

$$f(u\sigma, x\sigma) \leftarrow_{EXT_{AC}(R)} f(f(s\sigma, t\sigma), x\sigma) \to_{R_{AC}} f(v, x\sigma),$$

then there is also a proper overlap $u\sigma \leftarrow_R f(s\sigma, t\sigma) \to_{R_{AC}} v$. The former overlap can be eliminated, if the latter can be eliminated.

The correctness of Peterson and Stickel's associative-commutative completion procedure follows from Proposition 3.25. A number of convergent systems have been derived with this procedure (e.g., Hullot 1980).

Example 3.26. Let E be the set of axioms for *Abelian groups*:

$x + 0$	$\approx$	x
$x + (-x)$	$\approx$	0
$x + (y + z)$	$\approx$	$(x + y) + z$
$x + y$	$\approx$	$y + x$

and let AC be the set of the associativity and commutativity axioms for $+$ and R be the set of rules

$$
\begin{array}{rcl}
x + 0 & \rightarrow & x \\
-0 & \rightarrow & 0 \\
--x & \rightarrow & x \\
-(x + y) & \rightarrow & -x + -y \\
x + -x & \rightarrow & 0 \\
(x + -x) + y & \rightarrow & y
\end{array}
$$

The rewrite system R_{AC} is convergent modulo AC.

Example 3.27. Augment the above set AC by the associativity and commutativity axioms for $*$, and augment R by rewrite rules

$$
\begin{array}{rcl}
x * 0 & \rightarrow & 0 \\
x * 1 & \rightarrow & x \\
x * -y & \rightarrow & -(x * y) \\
(x * -y) * z & \rightarrow & -(x * y) * z \\
x * (y + z) & \rightarrow & x * y + x * z \\
(x * (y + z)) * x' & \rightarrow & (x * y) * x' + (x * z) * x'
\end{array}
$$

Then R_{AC} is an AC-convergent system for *associative-commutative rings with unit*.

Example 3.28. Hsiang (1985) has presented a rewrite system

$$
\begin{array}{rcl}
x \oplus 0 & \rightarrow & x \\
x \oplus x & \rightarrow & 0 \\
x \wedge 0 & \rightarrow & 0 \\
x \wedge 1 & \rightarrow & x \\
x \wedge x & \rightarrow & x \\
-x & \rightarrow & x \\
x \wedge (y \oplus z) & \rightarrow & x \wedge y \oplus x \wedge z
\end{array}
$$

where $\oplus$ (exclusive disjunction) and $\wedge$ (conjunction) are AC operators. The corresponding system R^e_{AC} constitutes a convergent system for *Boolean rings*.

Termination of associative-commutative rewrite systems can be proved by polynomial orderings (Lankford 1975, Ben Cherifa and Lescanne 1987) or by associative path orderings (Bachmair and Plaisted 1985; Bachmair and Dershowitz 1986). Termination proofs for the above systems can be found in Peterson and Stickel (1981) and Hsiang (1985).

3.7. The Protected Rule Method

Jouannaud and Kirchner (1986) described a completion procedure that uses both extended rules and critical pairs of $CP_A(R, A)$. We shall discuss their approach within our formalism. We assume that the proper subterm ordering modulo A is well-founded. (This implies that A contains no equation $t \simeq x$ and, by Lemma 3.11, that the encompassment ordering modulo A is well-founded.) We denote by $\mathcal{J}_A^{\succ}$ the inference system $\mathcal{E}_A^{\succ}$, with the collapse rule replaced by the following inference rule:

$$\text{COLLAPSE:} \qquad \frac{E; N \cup \{s \to t\}; S}{E \cup \{u \approx t\}; N; S} \qquad \begin{array}{l} \text{if } s \leftrightarrow_A^{\leq p} s' \to_{v \approx_n w}^p u, \\ \text{where either } p \neq \lambda \text{ or} \\ \text{else } s = s' \text{ and } s \rhd_A v \end{array}$$

Let $\mathcal{R}'_{\mathcal{J}}$ be the proof transformation system obtained from $\mathcal{R}_{\mathcal{E}}$ by changing the transformation rules reflecting collapse accordingly. We will further augment $\mathcal{R}'_{\mathcal{J}}$ by transformation rules for elimination of cliffs. The Extended Critical Pair Lemma suggests a transformation rule

$$s \leftrightarrow_A u \to_{R_A} t \quad \Rightarrow \quad s \leftrightarrow_A^* v \leftrightarrow_E w \leftrightarrow_A^* t$$

for elimination of proper overlaps. The difficulty with this transformation rule is that when we add it to $\mathcal{R}'_{\mathcal{J}}$ the resulting proof transformation system is not terminating. Therefore we shall use a slightly more restrictive version

$$s \leftrightarrow_{w \approx_0 v}^p u \leftrightarrow_A^{\leq q} u' \to_{v' \approx_n w'}^q t \quad \Rightarrow \quad s \leftrightarrow_A^{\leq p} s' \to_{v'' \approx_\perp w''}^p t' \leftrightarrow_A^* t,$$

where (i) $v' \succ w'$, $v'' \succ w''$, (ii) $n = \perp$ or $n = 1$, and (iii) the position q is strictly below p.

This transformation rule is applicable when an A-critical pair is computed and then oriented, before any simplifications are performed. Thereby the range of possible simplifications is restricted, as the inference rules of composition and collapse together are less powerful than simplification of unoriented equations.

By $\mathcal{R}_{\mathcal{J}}^{\succ}$ we denote the transformation system $\mathcal{R}'_{\mathcal{J}}$, augmented by the transformation rules for elimination of proper overlaps. The corresponding transformation relation is denoted by $\Rightarrow_{\mathcal{J}}$.

Lemma 3.29. *If the proper subterm ordering modulo A is well-founded, then the transformation system $\mathcal{R}_{\mathcal{J}}$ is terminating.*

Proof. Let P be a proof

$$t_0 \leftrightarrow_{e_1}^{p_1} t_1 \leftrightarrow_{e_2}^{p_2} t_2 \leftrightarrow_{e_3}^{p_3} \cdots t_{n-1} \leftrightarrow_{e_n}^{p_n} t_n$$

where e_i is an equation $u_i \approx_{n_i} v_i$. We redefine the complexity of single proof steps as follows:

$$
\begin{cases}
(\{[t_{i-1}], [t_i]\}, \bot, \bot, \bot, \bot) & \text{if } n_i = \top \\
(\{[t_{i-1}]\}, [t_{i-1}|_{p_i}], \beta_i^P, \bot, [t_i]) & \text{if } n_i = 0 \\
(\{[t_{i-1}]\}, [t_{i-1}|_{p_i}], \bot, u_i, [t_i]) & \text{if } n_i = \bot \text{ and } u_i \succ v_i \\
(\{[t_i]\}, [t_i|_{p_i}], \bot, v_i, [t_{i-1}]) & \text{if } n_i = \bot \text{ and } v_i \succ u_i \\
(\{[t_{i-1}]\}, \bot, |p_i|, \bot, [t_i]) & \text{if } n_i = 1 \text{ and } u_i \succ v_i \\
(\{[t_i]\}, \bot, |p_i|, \bot, [t_{i-1}]) & \text{if } n_i = 1 \text{ and } v_i \succ u_i
\end{cases}
$$

The complexity of a proof is the multiset of all tuples corresponding to its proof steps.

Complexity tuples are compared lexicographically by the multiset extension $\succ_{mul}$ of the reduction ordering $\succ$ (in the first component), the subterm ordering modulo A (in the second component), the greater-than relation (in the third component), the standard encompassment ordering (in the fourth component), and the reduction ordering $\succ$ (in the last component). We denote this ordering by $\succ^j$. Proofs are compared according to their complexity, using the multiset extension of $\succ^j$. We denote this ordering by $\succ_{\mathcal{J}}$. It is well-founded, if the subterm ordering modulo A is well-founded, and contains all transformation rules of $\mathcal{R}_{\mathcal{J}}$.

The above complexity measure differs from the one used in the proof of Lemma 3.19 mainly in that proof steps by equations in A have higher complexity. Most transformation rules can be shown to be decreasing by similar arguments as in the proof of Lemma 3.19. We consider the remaining cases in detail.

i) *Collapse.* Consider the transformation rule

$$
s \to^\lambda_{s \approx_\bot t} t \quad \Rightarrow \quad s \leftrightarrow^{\leq p}_A s' \to^p_{v \approx_n w} u \leftrightarrow^\lambda_{u \approx_\top t} t
$$

where $s \succ t$, $v \succ w$, $s \vartriangleright_A v$, and $n = \bot$ or $n = 1$. Moreover, either $p \neq \lambda$ or else $s = s'$.

If $p \neq \lambda$, then

$$
s \to^\lambda_{s \approx_\bot t} t \quad \succ_{\mathcal{J}} \quad s \leftrightarrow^{\leq p}_A s' \to^p_{v \approx_n w} u \leftrightarrow^\lambda_{u \approx_\top t} t
$$

(using the first two complexity components).

If $p = \lambda$, then $s = s'$ and $s \vartriangleright v$, so that

$$
s \to^\lambda_{s \approx_\bot t} t \quad \succ_{\mathcal{J}} \quad s \to^\lambda_{v \approx_n w} u \leftrightarrow^\lambda_{u \approx_\top t} t
$$

(using the first four complexity components).

ii) *Cliffs, variable overlaps.* Let P be a variable overlap $s \leftarrow^p_{w \approx_0 v} u \leftrightarrow^{\leq q}_A t' \to^q_{v' \approx_n w'} t$, where $q = pq'q''$ and $v|_{q'} = x$, for some variable x, and either

$n = \bot$ or $n = 1$. Let Q be the proof $s \leftarrow^P_{w \approx_0 v} u$ and Q' the proof

$$u|_{pq'} \leftrightarrow^{\leq q''}_A t'|_{pq'} \rightarrow^{q''}_{v' \approx_n w'} t|_{pq'}$$

so that $P = Qu[Q']_{pq'}$. As we have shown previously, the variable over-lap P can be replaced by a rewrite proof modulo A of the form $P' = P_1 \cdots P_k Q'' Q_m \cdots Q_1$ (cf. the proof of Lemma 3.19).

Since all terms in the proof $Q_m \cdots Q_1$ are strictly smaller than t, we have $u[Q']_{pq'} \succ_{\mathcal{J}} Q_m \cdots Q_1$. Furthermore $k > 0$, so that $Q \succ_{\mathcal{J}} P_1 \cdots P_k Q''$. (Recall that A contains no equations $t \approx x$. Therefore, all proof steps in $P_1 \cdots P_k$ apply at positions strictly below p.) We have $P \succ_{\mathcal{E}} P'$.

iii) *Cliffs, proper overlaps.* Consider a proper overlap

$$s \leftrightarrow^P_{w \approx_0 v} u \leftrightarrow^{\leq q}_A u' \rightarrow^q_{v' \approx_n w'} t$$

and the corresponding proof

$$s \leftrightarrow^{<p}_A s' \rightarrow^P_{v'' \approx_\bot w''} t' \leftrightarrow^*_A t,$$

where (i) $v' \succ w'$, $v'' \succ w''$, (ii) $n = \bot$ or $n = 1$, and (iii) the position q is strictly below p. The proof step $s \leftrightarrow^P_{w \approx_0 v} u$ is more complex in the second component than each proof step in $s \leftrightarrow^{<p}_A s'$. It is more complex than $s' \rightarrow^P_{v'' \approx_\bot w''} t'$ in the third component; and more complex than each proof step in $t' \leftrightarrow^*_A t$ in the first component.

Finally, in all cases that need to be considered $P \succ_{\mathcal{J}} P'$ also implies $Q[u[P\sigma]_p] \succ_{\mathcal{J}} Q[u[P'\sigma]_p]$.   Q.E.D.

We can now describe (non-failing) derivations that induce normalizing sequences of proof transformations.

Lemma 3.30. *A non-failing derivation $E_0; N_0; S_0 \vdash_{\mathcal{E}} E_1; N_1; S_1 \vdash_{\mathcal{E}} \cdots$ is fair with respect to $\mathcal{R}_{\mathcal{J}}$ and $\mathcal{N}_{\mathcal{E}}$ if (a) $CP_A(R_\infty)$ is a subset of $\bigcup_k E_k$, and (b) for each rule $s \rightarrow t$ in R_∞ and equation $u \approx v$ in A for which some non-variable subterm $u|_p$ is A-unifiable with s, either the extended rule $u[s]_p \rightarrow u[t]_p$ is contained in $\bigcup_k S_k$, or else all A-critical pairs of $s \rightarrow t$ on $u \approx v$ at position p are contained in $\bigcup_k R_k$.*

Proof. The proof is by induction on $\Rightarrow^+_{\mathcal{J}}$ and differs from the proof of Lemma 3.20 only in the elimination of proper overlaps $s \leftarrow_A u \rightarrow_{(R_\infty)_A} t$. The given assumptions ensure that each such cliff can be transformed by $\Rightarrow^+_{\mathcal{J}}$ to a simpler proof.   Q.E.D.

An A-completion procedure is said to be *fair* if all non-failing derivations it generates satisfy conditions (a) and (b) in the previous lemma.

Lemmas 3.29 and 3.30 indicate that Theorem 2.10 can be applied to the inference system $\mathcal{J}$.

Theorem 3.31. *Let A be a set of equations with a minimal complete uni-fication algorithm, such that the proper subterm ordering modulo A is well-founded. If an A-completion procedure is fair and does not fail for the given input, then $(N_\infty \cup S_\infty)_A$ is convergent modulo A.*

The theorem applies to all sets of equations A for which the proper sub-term ordering modulo A is well-founded. This excludes, for instance, theo-ries with identity axioms $f(x, e) \approx x$, or equipotency axioms $f(f(x)) \approx x$. Theories containing such axioms can be dealt with by the extended rule method. The theorem also indicates that A-critical pairs provide an alter-native to extended rules. The procedure given by Jouannaud and Kirchner (1986) is fair in our sense, but may on occasion require computation of extended (or at least protected) rules.

For instance, in the Jouannaud-Kirchner procedure, if in an A-critical overlap $v\sigma \leftrightarrow_A u\sigma \to_{R_A} u\sigma[t\sigma]$ of a rule $s \to t$ on an equation $u \approx v$, the term $v\sigma$ is not reducible by R, then the extended rule $u[s] \to u[t]$ is generated instead of the A-critical pair $v\sigma \approx u\sigma[t\sigma]$. Our definition of fairness covers A-completion procedures that never generate extended (or protected) rules. With such procedures it is possible to construct reduced systems.

An extended rewrite system R_A is said to be *reduced* if, for every rule $s \to t$ in R, (i) no proper subterm of s is reducible by R_A, (ii) the term s is not reducible by $R \setminus \{s \to t\}$, and (iii) t is not reducible by R_A.

Let R be the limit of a fair derivation in $\mathcal{J}_A$ in which no rules have been protected and neither composition nor collapse apply to any persisting rule. Then (i) no right-hand side of any rule in R is reducible by R_A (for otherwise a composition inference could have been applied), (ii) no proper subterm of a left-hand side is reducible by R_A, and (iii) no left-hand side is a proper instance of another left-hand side (for otherwise a collapse inference could have been applied). The system R_A, though convergent modulo A, need not be reduced as it may contain two distinct rules $s \to t$ and $u \to v$, such that $s \leftrightarrow^*_A u\sigma$, for some substitution σ. But we can apply the following proposition (cf. Jouannaud and Kirchner 1986) to obtain a reduced rewrite system from R.

Proposition 3.32. *Let R be a finite rewrite system and A be a set of equations, such that the proper subterm ordering modulo A is well-founded and R_A is convergent modulo A. Let R' be the system obtained from R by deleting any rule $s \to t$ for which there is a rule $u \to v$, distinct from $s \to t$, such that $s \leftrightarrow^*_A u\sigma$, for some substitution σ. Then R'_A is also convergent modulo A.*

The protection rule can be used to allow for a slightly more general version of the collapse inference rule. Suppose $s \approx t$ is an A-critical pair in $CP_A(N, A)$ and let $u \to v$ be a rule in R, such that $s \leftrightarrow^+_A u\sigma$, for some substitution σ. The equation $s \approx t$ can be simplified and replaced by $v\sigma \approx t$. However, fairness requires that $s \approx t$ be oriented into a rewrite rule $s \to t$. Since collapse does not allow reduction by R_A at the top, the rule $s \to t$ cannot be collapsed. However, instead of adding $s \to t$ as a rule, we can add the equation $v\sigma \approx t$, while at the same time protecting the rule $u \to v$. This inference is reflected by transformation rules

$$s \leftrightarrow^p_{w \approx_0 v} u \leftrightarrow^{\leq q}_A u' \to^q_{v' \approx_n w'} t$$
$$\Rightarrow$$
$$s \leftrightarrow^{<p'}_A s' \to^{p'}_{v'' \approx_1 w''} s'' \leftrightarrow^{p'}_{v''' \approx_\top w'''} t' \leftrightarrow^*_A t,$$

where (i) $v' \succ w'$, $v'' \succ w''$, (ii) $n = \bot$ or $n = 1$, and (iii) the position q is strictly below p and p' is below p. This proof transformation is simplifying with respect to $\succ_J$, as all proof steps in $s \leftrightarrow^{<p'}_A s'$ are bound on the right, while $s \leftrightarrow^p_{w \approx_0 v} u$ is not. Therefore, the above correctness result also holds if we extend the inference system $\mathcal{E}$ by a corresponding new inference rule. Jouannaud and Kirchner (1986) implicitly use this kind of inference. The disadvantage is that the rule $u \to v$ has to be protected, which in the final analysis may increase the number of rules and prevent construction of reduced systems.

In their completion procedure, Jouannaud and Kirchner also consider rewrite relations $L \cup N_A$, where L contains only left-linear rewrite rules, and combine extended rewriting with standard rewriting in an attempt to avoid the costly operations of A-matching and A-unification when dealing with left-linear rules. This refinement can readily be formalized within our framework. The goal of such a modified completion procedure is to construct a rewrite system $R = L \cup N \cup S$, such that $L \cup (N \cup S)_A$ is convergent modulo A. In other words, this means that all proofs can be transformed to rewrite proofs modulo A in $L \cup (N \cup S)_A$. This requires elimination of cliffs $s \leftarrow_L u \leftrightarrow_A t$, where the second proof step is below the first. Such cliffs correspond to extended rewrite steps $s \leftarrow_{L_A} t$, but the extended rewrite system L_A is not used. Elimination of such cliffs can be achieved by standard critical pair computation. The proof transformation relation $\Rightarrow_J$ can be extended accordingly; for a brief discussion see Bachmair and Dershowitz (1989, p. 198).

3.8. Extended Critical Pair Criteria

Just as with standard completion, the efficiency of A-completion can often be improved by sifting out redundant A-critical pairs. Critical pair criteria for associative-commutative completion have been discussed by Winkler

(1984), Küchlin (1986a), and Kapur, Musser, and Narendran (1988). We shall briefly describe how the concept of blocking (Slagle 1974) can be adapted to A-completion, for sets A for which the proper subterm ordering modulo A is well-founded.

A proof step $s \to_{u \approx v}^p t$ is said to be *blocked* with respect to R and A if, for all variables x, the term $x\sigma$ is irreducible by R/A, where σ is a substitution, such that $s|_p = u\sigma$ and $t|_p = v\sigma$.

For example, if R contains rules $-(x + y) \to -x + -y$ and $0 + x \to x$, where $+$ is an AC operator, then the rewrite step $-((z + 0) + y) \to_R -(z + 0) + -y$ is not blocked with respect to R and AC, because the term $z + 0$, which is substituted for x, is reducible by R/AC. Most non-blocked rewrite steps can be eliminated from proofs.

Let P be a proof step $s \to_{u \approx_n v}^p t$, where u is not a variable, and let σ be a substitution, such that $s|_p = u\sigma$ and $t|_p = v\sigma$. Suppose u contains a variable x that is reducible by R/A, say $x\sigma \to_{R/A} w$. Then P can be replaced by a proof

$$Q = s \to_{R/A}^* s[u\tau]_p \to_{u \approx_n v}^p t[v\tau]_p \leftarrow_{R/A}^* t$$

where $x\tau = w$ and $y\tau = y\sigma$, if $y \neq x$. If $u \approx_n v$ is an unprotected rule (i.e., $n = \bot$), then $P \succ_{\mathcal{J}} Q$.

An A-critical overlap $s \leftarrow_R u \to_{R_A} t$ is said to be *blocked* if its first and last step are blocked; an A-critical overlap $s \leftrightarrow_A u \to_{R_A} t$ is blocked if its last step is blocked. An A-critical pair is blocked if its corresponding A-critical overlap is. Non-blocked A-critical overlaps that involve unprotected rules only, define redundant A-critical pairs.

In the case of associative-commutative completion, a restricted form of blocking can be applied to extended (i.e., protected) rules. For instance, proof steps $f(s\sigma, x\sigma) \to_R f(t\sigma, x\sigma)$ in which an extended rule $f(s, x) \to f(t, x)$ is applied with a substitution σ, such that $y\sigma$ is reducible by R/AC, for some variable y in s, can be simplified by the above proof transformation. Informally, proof steps that "extend" a non-blocked rewrite step $s\sigma \to_R t\sigma$ are redundant. The "extension variable" x, on the other hand may well be instantiated by a reducible term.

For example, consider the two rules, $a + b \to c$ and $(a + a) + (b + b) \to d$, where the operator $+$ is associative-commutative. The only AC-critical pairs are those involving the extended rules, $(a + b) + x \to c + x$ and $((a + a) + (b + b)) + x \to d + x$. They require that the extension variable x be instantiated by a reducible term. If those AC-critical pairs are considered redundant, no new equation is generated. However, the system is not convergent modulo AC, as $(a + b) + (a + b)$ has two different normal-forms, $c + c$ and d, that are not equivalent with respect to AC.

Summary

We have formalized extended completion as an equational inference system and have established the correctness of various extended completion procedures. The left-linear rule method proposed by Huet (1980) uses standard matching and unification, but is limited to left-linear rewrite systems, while other methods that also apply to non-left-linear rules, require A-matching and A-unification.

The extended rule method is characterized by the systematic use of extended rules. We have derived sufficient conditions for its fairness and have shown that the fairness conditions can be considerably weakened in the case of associativity and commutativity. Our correctness result covers, for instance, the associative-commutative completion procedure described by Peterson and Stickel (1981). More recently, Baird, Peterson, and Wilkerson (1989) and Jouannaud and Marché (1990) have described completion procedures for sets A of associativity, commutativity, and identity axioms that also fall into this general framework. The protected rule method generalizes the Jouannaud-Kirchner procedure. We have proved its correctness under the assumption that the proper subterm ordering modulo A is well-founded, which covers all theories A for which congruence classes are finite.

4. ORDERED COMPLETION

Standard completion fails whenever an equation $s \approx t$ is generated, such that s and t are irreducible, yet incomparable with respect to the given reduction ordering. Examples of such unorientable equations are commutativity axioms $x \cdot y \approx y \cdot x$, as the two terms $x \cdot y$ and $y \cdot x$ are incomparable with respect to any reduction ordering.

For example, let E_0 consist of the equations $a \approx b$, $a \approx c$, $fb \approx b$, and $fa \approx d$, and let $\succ$ be the lexicographic path ordering corresponding to a precedence $\succ_p$ in which $a \succ_p b \succ_p d$ and $a \succ_p c \succ_p d$. The derivation

$$E_0 \, ; \emptyset \quad \vdash_C^* \quad \{b \approx c, fb \approx b, fb \approx d\} \, ; \{a \to b\}$$
$$\vdash_C^* \quad \{b \approx c, b \approx d\} \, ; \{a \to b, fb \to b\}$$
$$\vdash_C^* \quad \{d \approx c, fd \approx d\} \, ; \{a \to d, b \to d\}$$
$$\vdash_C^* \quad \emptyset \, ; \{a \to d, b \to d, c \to d, fd \to d\}$$

yields a convergent system, whereas the alternative derivation

$$E_0 \, ; \emptyset \quad \vdash_C^* \quad \{b \approx c, fb \approx b, fc \approx d\} \, ; \{a \to c\}$$
$$\vdash_C \quad \{b \approx c, fb \approx b\} \, ; \{a \to c, fc \to d\}$$
$$\vdash_C \quad \{b \approx c\} \, ; \{a \to c, fb \to b, fc \to d\}$$

fails (Dershowitz, Marcus, and Tarlecki 1988).

This example indicates that the order in which inference rules are applied may determine whether or not an unorientable equation is generated.[1] To avoid failure a completion procedure may have to systematically enumerate all possible derivations, e.g., via backtracking. In some cases, standard completion is bound to fail even with backtracking. In fact, the method may fail even when a convergent system does exist and it is supplied with a suitable reduction ordering.

For example, let E_0 be the set of equations

$$
\begin{array}{rcl}
1 * (-x + x) & \approx & 0 \\
1 * (x + -x) & \approx & x + -x \\
-x + x & \approx & y + -y
\end{array}
$$

---

[1] It might be argued that in the above example the ordering could be extended so that b and c are comparable. However, one can easily devise a similar example in which b and c are replaced by terms (with variables) that are intrinsically incomparable.

and $R_0 = \emptyset$. Standard completion fails, regardless of which reduction ordering is supplied as input! The only inference rule that can possibly be applied to the above set of equations is orientation, which may result in two rules

$$
\begin{aligned}
1 * (-x + x) &\;\rightarrow\; 0 \\
1 * (x + -x) &\;\rightarrow\; x + -x
\end{aligned}
$$

(no other orientation is possible). These two rules do not overlap and form a convergent rewrite system. The third equation is unorientable and both sides are irreducible by the above two rules. As a consequence, no further inference rules are applicable. Nonetheless, there exists a convergent system

$$
\begin{aligned}
-x + x &\;\rightarrow\; 0 \\
x + -x &\;\rightarrow\; 0 \\
1 * 0 &\;\rightarrow\; 0
\end{aligned}
$$

for the given equational theory.

Unorientable equations in which one side contains a variable not occurring in the other side can sometimes be dealt with by introducing new function symbols, a technique suggested by Knuth and Bendix (1967). Suppose, for instance, that the equation $-x + x \approx y + -y$ is replaced by two rules $-x + x \rightarrow c$ and $y + -y \rightarrow c$, where c is a new (minimal) constant. Completion would then succeed in constructing a convergent system of four rules

$$
\begin{aligned}
-x + x &\;\rightarrow\; c \\
x + -x &\;\rightarrow\; c \\
1 * c &\;\rightarrow\; c \\
0 &\;\rightarrow\; c
\end{aligned}
$$

which represents a conservative extension of the original equational theory (and hence also provides a decision procedure). But very often this technique leads to the introduction of ever more new function symbols, and the above example does indicate an inherent inadequacy of standard completion. In general, given a reduction ordering $\succ$, a set of equations can be represented by a convergent rewrite system contained in $\succ$ if and only if every congruence class of $\leftrightarrow^*_E$ has a unique minimum element with respect to $\succ$ (Avenhaus 1986).

In this chapter, we shall pursue a different approach for dealing with unorientable equations, called ordered completion, which is designed to construct sets of equations that define unique ground normal forms. This ordered completion method is guaranteed to find a convergent system, if one exists and the reduction ordering supplied to the procedure satisfies some reasonable conditions. It requires neither backtracking nor introduction of new function symbols and is a refutationally complete theorem prover for equational theories, but has the advantage over paramodulation (Robinson and Wos 1969) in that equations can always be kept in fully simplified form and fewer equational consequences need to be considered, as the ordering supplied to the procedure gives some measure of direction to the prover. The method works with all general purpose orderings that have been proposed for rewriting, e.g., polynomial interpretations and recursive path orderings. In the more general context of Horn clauses with equality it yields an inference system consisting of ordered versions of positive unit resolution and paramodulation (with simplification).

4.1.   Ordered Completion

A convergent rewrite system defines unique normal forms for *all* terms, while in many applications, including (refutational) theorem proving, the uniqueness of *ground* normal forms may be sufficient. In this section we will present an inference system, called ordered completion, for the construction of sets of equations that define unique ground normal forms.

We first refine the notion of rewriting, so that all equations (and not just rewrite rules) can potentially be used for simplification. For example, even though the commutativity axiom $x \cdot y \approx y \cdot x$ cannot be oriented with respect to any reduction ordering, some of its *instances* may be orientable (e.g., we have $f(x) \cdot x \succ_{lpo} x \cdot f(x)$ for the lexicographic path ordering).

Definition 4.1. We call $u\sigma \approx v\sigma$ an *orientable instance* (with respect to a reduction ordering $\succ$) of an equation $u \simeq v$ if $u\sigma \succ v\sigma$. By $E^{\succ}$ we denote the set of all orientable instances of equations in E.

Evidently, $s \rightarrow_{E^{\succ}} t$ if and only if $s \leftrightarrow_E t$ by an equation $u \simeq v$ and a substitution σ, such that $u\sigma \succ v\sigma$. In particular, $s \rightarrow_{E^{\succ}} t$ implies $s \leftrightarrow_E t$ and $s \succ t$ (but not vice versa). By definition, the rewrite relation $\rightarrow_{E^{\succ}}$ induced by $E^{\succ}$ is terminating. Also note that every instance of an orientable instance is orientable. Rewrite rules can be thought of as equations all instances of which are orientable in the same direction.

Definition 4.2. A set of equations E is said to be *ground convergent* (with respect to a reduction ordering $\succ$) if for all ground terms s and t with $s \leftrightarrow_E^* t$, there exists a ground term v, such that $s \rightarrow_{E^{\succ}}^* v \leftarrow_{E^{\succ}}^* t$.

A ground convergent system defines unique ground normal forms.

In formulating ordered completion as an equational inference system we take a different approach than with standard completion and describe the inference rules in terms of a proof ordering. We shall consider arbitrary sets of labelled equations. (Recall that the label of an equation is either a non-negative number or one of the symbols $\top$ or $\bot$.)

Let $\succ$ be a reduction ordering. The cost of a single equational proof step $s \leftrightarrow^p_{u \approx_n v} t$ is defined to be

$$\left\{ \begin{array}{ll} (\{s\}, u, n, t) & \text{if } s \succ t \\ (\{t\}, u, n, s) & \text{if } t \succ s \\ (\{s, t\}, \bot, \bot, \bot) & \text{otherwise.} \end{array} \right.$$

The complexity of a proof is the multiset of the costs of its proof steps. By $\succ^c$ we denote the lexicographic combination of (i) the multiset extension of the reduction ordering $\succ$, (ii) the encompassment ordering $\succeq$, (iii) the ordering on labels, and (iv) the reduction ordering $\succ$. The ordering $\succ^c$ is used to compare tuples representing the cost of single proof steps; proofs are compared according to their complexity, using the multiset extension of $\succ^c$. We denote this ordering on proofs by $\succ_\mathcal{O}$. It is indeed a proof reduction ordering.

The inference system $\mathcal{O}^\succ$ (also denoted by $\mathcal{O}$ if $\succ$ is clear from the context) consists of two inference rules:

DEDUCTION: $$\frac{E}{E \cup \{s \approx_n t\}} \qquad \text{if } s \leftrightarrow^*_E t$$

DELETION: $$\frac{E \cup \{s \approx_n t\}}{E} \qquad \begin{array}{l} \text{if there is a proof } P \text{ of} \\ s \approx t \text{ in } E \text{ such that} \\ s \leftrightarrow^\lambda_{s \approx_n t} t \succ_\mathcal{O} P \end{array}$$

where E denotes a finite set of (labelled) equations. Inference systems $\mathcal{O}^\succ$ are called *ordered completion systems*.

Ordered completion systems are evidently sound.

Lemma 4.3. *Whenever $E \vdash_\mathcal{O} E'$, then the two congruence relations $\leftrightarrow^*_E$ and $\leftrightarrow^*_{E'}$ are the same.*

Moreover, the proof reduction ordering $\succ_\mathcal{O}$ trivially reflects the inference system $\mathcal{O}^\succ$.

For practical purposes, the inference rules are not needed in the generality in which we have formulated them above and implementations are usually based on derived inference rules. Simplification techniques, in particular, can be described as certain combinations of deduction and deletion inferences. As a matter of fact, ordered completion generalizes standard completion, in that all inference rules of standard completion are derived inference rules of ordered completion.

Lemma 4.4. *If $E \vdash_C E'$, then $E \vdash_O^* E'$.*

Proof. The various inference rules of standard completion can be derived from ordered completion as follows.

Deduction. The inference rule for standard completion is a special case of the corresponding inference rule of ordered completion.

Orientation. Suppose we have $E \cup \{s \approx_\top t\} \vdash_C E \cup \{s \approx_\bot t\}$, where $s \succ t$. Then there is a derivation by ordered completion,

$$E \cup \{s \approx_\top t\} \vdash_O E \cup \{s \approx_\top t, s \approx_\bot t\} \vdash_O E \cup \{s \approx_\bot t\},$$

where the first inference is by deduction and the second by deletion. (Note that $s \to^\lambda_{s \approx_\top t} t \succ_O s \to^\lambda_{s \approx_\bot t} t$.)

Deletion. Observe that $E \cup \{s \approx_n s\} \vdash_O E$ is a deletion inference by ordered completion, as the proof $s \leftrightarrow^\lambda_{s \approx_n s} s$ has greater complexity than the empty proof.

Simplification. Suppose $E \cup R \cup \{s \approx_\top t\} \vdash_C E \cup R \cup \{u \approx_\top t\}$ is a simplification inference in standard completion. Then there is a rewrite step $s \to^p_{v \approx_\bot w} u$ in R, where $v \succ w$ (and hence $s \succ u$). We have $u \leftrightarrow^*_{E \cup R} t$, so that there is a deduction inference

$$E \cup R \cup \{s \approx_\top t\} \vdash_O E \cup R \cup \{s \approx_\top t, u \approx_\top t\}$$

by ordered completion. Now compare the proof $s \leftrightarrow^\lambda_{s \approx_\top t} t$ with $s \to^p_{v \approx_\bot w} u \leftrightarrow^\lambda_{u \approx_\top t} t$. The cost of $s \leftrightarrow^\lambda_{s \approx_\top t} t$ is $(\{s\}, s, \top, t)$, if $s \succ t$; $(\{t\}, t, \top, s)$, if $t \succ s$; and $(\{s, t\}, \bot, \bot, \bot)$, otherwise. The cost of $s \to^p_{v \approx_\bot w} u$ is $(\{s\}, s, \bot, u)$. The cost of $u \leftrightarrow^\lambda_{u \approx_\top t} t$ is $(\{t\}, t, \top, u)$, if $t \succ u$; $(\{u\}, u, \top, t)$, if $u \succ t$; and $(\{u, t\}, \bot, \bot, \bot)$, otherwise. Checking all the different cases, we find that

$$s \leftrightarrow^\lambda_{s \approx_\top t} t \succ_O s \to^p_{v \approx_\bot w} u \leftrightarrow^\lambda_{u \approx_\top t} t,$$

so that there is a deletion inference,

$$E \cup R \cup \{s \approx_\top t, u \approx_\top t\} \vdash_O E \cup R \cup \{u \approx_\top t\}.$$

Composition. Suppose we have $E \cup R \cup \{s \approx_\bot t\} \vdash_C E \cup R \cup \{s \approx_\bot u\}$ and $t \to^p_{v \approx_\bot w} u$ is a proof step in R, where $s \succ t$ and $v \succ w$ (and thus $t \succ u$). There is a deduction inference

$$E \cup R \cup \{s \approx_\bot t\} \vdash_O E \cup R \cup \{s \approx_\bot t, s \approx_\bot u\}.$$

Furthermore, we have

$$s \to^\lambda_{s \approx_\bot t} t \succ_O s \to^\lambda_{s \approx_\bot u} u \leftarrow^p_{w \approx_\bot v} t$$

because $s \succ t \succ u$ implies that $(\{s\}, s, \perp, t)$ is more complex than both $(\{s\}, s, \perp, u)$ and $(\{t\}, v, \perp, u)$. Thus there is a deletion inference

$$E \cup R \cup \{s \approx_\perp t, s \approx_\perp u\} \vdash_\mathcal{O} E \cup R \cup \{s \approx_\perp u\}.$$

Collapse. Finally, let $E \cup R \cup \{s \approx_\perp t\} \vdash_C E \cup R \cup \{u \approx_\top t\}$ be a collapse inference. Then there is a proof step $s \to^p_{v \approx_\perp w} u$ in R, where $v \succ w$ (and thus $s \succ u$) and $s \triangleright v$. Moreover, we have $s \succ t$. There is a deduction inference

$$E \cup R \cup \{s \approx_\perp t\} \vdash_\mathcal{O} E \cup R \cup \{s \approx_\perp t, u \approx_\top t\}$$

by ordered completion. Moreover,

$$s \to^\lambda_{s \approx_\perp t} t \quad \succ_\mathcal{O} \quad s \to^p_{v \approx_\perp w} u \leftrightarrow^\lambda_{u \approx_\top t} t$$

because $s \succ u$ and $s \triangleright v$ imply that $(\{s\}, s, \perp, t)$ is more complex than both $(\{s\}, v, \perp, u)$ and $(\{u, t\}, \perp, \perp, \perp)$ (and the cost of $u \leftrightarrow^\lambda_{u \approx_\top t} t$ is at most $(\{u, t\}, \perp, \perp, \perp)$). Consequently there is a deletion inference

$$E \cup R \cup \{s \approx_\perp t, u \approx_\top t\} \vdash_\mathcal{O} E \cup R \cup \{u \approx_\top t\}$$

by ordered completion.   Q.E.D.

Definition 4.5. An *ordered completion procedure* is a program that takes as input a set of equations E_0 and a reduction ordering $\succ$, and uses the inference rules of $\mathcal{O}^\succ$ to generate derivations from E_0. Such a procedure is said to *succeed* for a set of equations E_0 and a reduction ordering $\succ$ if E_∞ is ground convergent with respect to $\succ$.

Consider, for example, an *entropic groupoid* defined by the two axioms

$$\begin{array}{rcl} (x \cdot y) \cdot (z \cdot w) & \approx & (x \cdot z) \cdot (y \cdot w) \\ (x \cdot y) \cdot x & \approx & x. \end{array}$$

The first equation is *permutative* and cannot be oriented in any reduction ordering. Standard completion will fail for this set of equations, whereas with ordered completion we can obtain a set of equations

$$\begin{array}{rcl} (x \cdot y) \cdot z & \approx & (x \cdot w) \cdot z \\ (x \cdot y) \cdot x & \to & x \\ x \cdot (y \cdot z) & \to & x \cdot z \\ ((x \cdot y) \cdot z) \cdot w & \to & x \cdot w \end{array}$$

(Hsiang and Rusinowitch 1987), which is ground convergent with respect to the lexicographic path ordering. (Orientable equations are written as rules.)

The set E therefore provides a decision procedure for the word problem in the above theory. Pedersen (1984) has designed another decision procedure based on a certain notion of extended rewriting.

We shall next derive sufficient conditions for the success of ordered completion.

Definition 4.6. A reduction ordering $\succ$ is said to be *complete* (with respect to E) if, whenever s and t are distinct ground terms (with $s \leftrightarrow^*_E t$), then either $s \succ t$ or $t \succ s$. An ordering is called *completable* if it can be extended to a complete reduction ordering.

Complete ordering are thus total on ground terms. We also say more specifically that $\succ$ can be completed to $>$ if $>$ is a complete reduction ordering containing $\succ$.

By a *ground rewrite proof* (with respect to $\succ$) in E we mean a proof of the form $s \to^*_{E\succ} v \leftarrow^*_{E\succ} t$. By definition, a set of equations E is ground convergent if and only if there exists a ground rewrite proof for every equation between ground terms that is provable in E. If the reduction ordering $\succ$ is complete, then each proof step $s \leftrightarrow_E t$, where $s \neq t$, is of the form $s \to_{E\succ} t$ or $s \leftarrow_{E\succ} t$. Thus, a ground rewrite proof with respect to a complete reduction ordering $\succ$ is simply a (ground) proof containing no peak $s \leftarrow_{E\succ} u \to_{E\succ} t$. Proof normalization corresponds to elimination of peaks $s \leftarrow_{E\succ} u \to_{E\succ} t$, and can be achieved by computation of suitable critical pairs.

Definition 4.7. Let $s \approx t$ and $u \approx v$ be two equations with no variables in common (the variables in one equation are renamed, if necessary) and suppose some non-variable subterm $s|_p$ of s is unifiable with u, σ being the most general unifier. We say that the *superposition* of $u \approx v$ on $s \approx t$ at position p determines an *ordered critical pair* $t\sigma \approx s\sigma[v\sigma]_p$ (with respect to the ordering $\succ$) if there exists a (ground) substitution τ, such that $s\sigma\tau \succ t\sigma\tau$ and $u\sigma\tau \succ v\sigma\tau$.

As before, the proof $t\sigma \leftarrow^\lambda_{t \approx s} s\sigma \to^p_{u \approx v} s\sigma[v\sigma]_p$ is called a *critical overlap*; the term $s\sigma$, the *overlapped term*; and the position p, the *critical pair position*. By $CP^\succ(E)$ we denote the set of all ordered critical pairs (with respect to $\succ$) between equations in $E \cup E^{-1}$.

For example, the two equations $(x \cdot y) \cdot (z \cdot w) \approx (x \cdot z) \cdot (y \cdot w)$ and $(x \cdot y) \cdot x \approx x$ overlap in $((u \cdot v) \cdot u) \cdot (v' \cdot v) \leftrightarrow_E ((u \cdot v) \cdot v') \cdot (u \cdot v) \leftrightarrow_E u \cdot v$ and define an ordered critical pair $((u \cdot v) \cdot u) \cdot (v' \cdot v) \approx u \cdot v$ with respect to the lexicographic path ordering.

An ordered critical pair of an equation on itself at the top need not be trivial. For instance, superposing the equation $x \cdot a \approx a^-$ on itself at the top may yield a non-trivial equation $x \cdot a \approx y \cdot a$.

The computation of ordered critical pairs with respect to a reduction ordering $\succ$ requires that one decide, given terms s, t, u, and v, whether there exists a ground substitution σ, such that $s\sigma \succ t\sigma$ and $u\sigma \succ v\sigma$. This question is decidable, for instance, for lexicographic path orderings based on a total precedence (Comon 1990). If such inequations cannot be solved for a given ordering $\succ$, then completion may have to deduce more equations than are actually necessary to ensure fairness. In any case, computation of standard critical pairs suffices for fairness, as the set of ordered critical pairs $CP^\succ(E)$ is a subset of $CP(E \cup E^{-1})$.

The Critical Pair Lemma 2.13 can be adapted to ordered critical pairs without much difficulty.

Lemma 4.8. (Ordered Critical Pair Lemma, Lankford 1975) *Let $\succ$ be a complete reduction ordering with respect to E. For all ground terms s, t, and u with $s \leftarrow_{E^\succ} u \rightarrow_{E^\succ} t$, there either exists a ground term v, such that $s \rightarrow^*_{E^\succ} v \leftarrow^*_{E^\succ} t$, or else $s \leftrightarrow_{CP^\succ(E)} t$.*

Proof. The proof is similar to the proof of the Critical Pair Lemma. Non-overlaps and proper overlaps, in particular, can be dealt with in the same way. The only difference is in the variable overlap case, where the fact that $\succ$ is a complete reduction ordering is needed to obtain a ground rewrite proof in $E^\succ$. More precisely, any variable overlap $s \leftarrow_{E^\succ} u \rightarrow_{E^\succ} t$ can be replaced by a proof of the form $s \rightarrow^*_{E^\succ} v \leftrightarrow_E w \leftarrow^*_{E^\succ} t$ (cf. the proof of the Critical Pair Lemma). This proof is a ground rewrite proof, as the completeness of $\succ$ implies that the proof step $v \leftrightarrow_E w$ is either of the form $v \rightarrow_{E^\succ} w$ or $v \leftarrow_{E^\succ} w$. Q.E.D.

The lemma suggests the following notion of fairness for ordered completion.

Definition 4.9. A derivation $E_0 \vdash_{O^\succ} E_1 \vdash_{O^\succ} \cdots$ is said to be *fair* if for every ground proof P of the form $s \leftarrow_{E_\infty^\succ} u \rightarrow_{E_\infty^\succ} t$, there exists a proof Q of $s \approx t$ in $\bigcup_i E_i$, such that $P \succ_O Q$.

Observe that whenever P is a proof $s \leftarrow_{E^\succ} u \rightarrow_{E^\succ} t$ and Q a corresponding rewrite proof $s \rightarrow^*_{E^\succ} v \leftarrow^*_{E^\succ} t$, then $P \succ_O Q$. Computation of ordered critical pairs between persisting equations ensures fairness.

Lemma 4.10. *Let $\succ$ be a complete reduction ordering with respect to E. A derivation by ordered completion from the initial set E is fair if the set of ordered critical pairs $CP^\succ(E_\infty)$ is a subset of the set of all derived equations $\bigcup_k E_k$.*

The completeness of the given reduction ordering can also be used to advantage, as suggested by Martin and Nipkow (1990). For example, let E

consist of the equations

$$(x \cdot y) \cdot z \ \approx \ x \cdot (y \cdot z)$$
$$x \cdot y \ \approx \ y \cdot x$$
$$x \cdot (y \cdot z) \ \approx \ y \cdot (x \cdot z)$$

and let $\succ$ be the lexicographic path ordering (which is complete in this case, as shown below). Then

$$z \cdot (x \cdot y) \leftrightarrow_E (x \cdot y) \cdot z \leftrightarrow_E x \cdot (y \cdot z)$$

is a critical overlap. Fairness requires that every ground instance

$$u \cdot (s \cdot t) \leftarrow_{E \succ} (s \cdot t) \cdot u \rightarrow_{E \succ} s \cdot (t \cdot u)$$

of this critical overlap can be replaced by a simpler proof. Since the ordering $\succ$ is complete, all ground terms are comparable. This additional knowledge may be exploited in constructing simpler proofs for the various ground instances of a given peak. For example, if $s \succ u \succ t$, then there exists a ground rewrite proof

$$u \cdot (s \cdot t) \rightarrow_{E \succ} u \cdot (t \cdot s) \rightarrow_{E \succ} t \cdot (u \cdot s) \leftarrow_{E \succ} t \cdot (s \cdot u) \leftarrow_{E \succ} s \cdot (t \cdot u).$$

On the other hand, if $s = u$ and $u \succ t$, then the peak can be replaced by a one-step proof

$$s \cdot (s \cdot t) \rightarrow_{E \succ} s \cdot (t \cdot s).$$

The above set of equations actually forms a ground convergent system for associative-commutative operators with respect to the lexicographic ordering (Martin and Nipkow 1990).

Fair derivations always succeed:

Theorem 4.11. *Let $\succ$ be a complete reduction ordering with respect to E and let E_∞ be the limit of a fair derivation from E by the ordered completion system $\mathcal{O}^\succ$. Then E_∞ is ground convergent with respect to $\succ$.*

Proof. Let $E_0 \vdash_{\mathcal{O}\succ} E_1 \vdash_{\mathcal{O}\succ} \cdots$ be a fair derivation and $\succ$ be a complete reduction ordering with respect to E_0. We will show that for every ground proof P of $s \approx t$ in $\bigcup_i E_i$, there exists a ground rewrite proof Q of $s \approx t$ in E_∞, such that $P \succeq_O Q$.

Suppose the assertion is not true. Then there exists some ground proof in $\bigcup_i E_i$ which is minimal with respect to the proof ordering $\succ_O$ but is not a ground rewrite proof. Let P be a minimal such proof. Since the proof ordering $\succ_O$ reflects ordered completion, P has to be a proof in E_∞.

Using the fact that the reduction ordering $\succ$ is complete and P is a minimal non-rewrite proof of $s \approx t$, we may infer that P actually has to be a peak $s \leftarrow_{E_\infty^\succeq} u \rightarrow_{E_\infty^\succeq} t$. But this contradicts the fairness assumption, which states that no peak in $E_\infty^\succeq$ is a minimal proof.   Q.E.D.

The theorem applies to complete reduction orderings. If a precedence ordering on function symbols is total, then it can be extended to a complete reduction ordering on terms by way of a lexicographic path ordering. In other words, lexicographic path orderings are completable.

Lemma 4.12. *Let $\succ_p$ be a total precedence ordering on function symbols. Then the corresponding lexicographic path ordering $\succ$ is complete.*

Proof. Suppose the lexicographic path ordering ordering $\succ$ is not complete. Let $s = f(s_1, \ldots, s_m)$ and $t = g(t_1, \ldots, t_n)$ be two distinct ground terms of minimal combined size, such that neither $s \succ t$ nor $t \succ s$.

Then the term s has to be comparable with all subterms t_i of t, and t has to be comparable with all subterms s_j of s. In fact, we must have $s \succ t_i$, for all i with $1 \leq i \leq n$, for otherwise $t \succ s$. By the same argument, $t \succ s_i$, for all i with $1 \leq i \leq m$. But then $f \succ_p g$ would imply $s \succ t$, and $g \succ_p f$ would imply $t \succ s$. Since $\succ_p$ is total, the two symbols f and g have to be identical.

Since s and t are distinct, we know that there exists an i with $1 \leq i \leq m$, such that $s_i \neq t_i$, while $s_j = t_j$, for all j with $j < i$. The two terms s_i and t_i must be comparable. Yet $s_i \succ t_i$ implies $s \succ t$ and $t_i \succ s_i$ implies $t \succ s$, either of which contradicts the assumption that s and t are incomparable with respect to $\succ$.   Q.E.D.

All general-purpose term orderings used in practice are completable. For instance, any ordering based on polynomial interpretations (Lankford 1975, 1979) can be extended to a complete ordering by combining it with a well-founded ordering to distinguish ground terms having the same interpretations.

Ordered completion is sound in that its inference rules preserve the given equational theory. If one is primarily interested in the *ground* equational theory, i.e., the set of all ground equations $s \approx t$ for which $s \leftrightarrow_E^* t$, then the following inference rules, which preserve only the ground theory, may be used for the construction of ground convergent systems:

$$\text{DEDUCTION:} \qquad \frac{E}{E \cup \{s \approx_n t\}} \qquad \text{if } s\sigma \leftrightarrow^{*}_{E} t\sigma \text{ for every ground substitution } \sigma$$

$$\text{DELETION:} \qquad \frac{E \cup \{s \approx_n t\}}{E}$$

if, for every ground substitution σ, there exists a proof P of $s\sigma \approx t\sigma$ in E such that $s\sigma \leftrightarrow^{\lambda}_{s \approx_n t} t\sigma \succ_{\mathcal{O}} P$

A specific inference system consisting of inference rules derived from these deduction and deletion rules has been described by Martin and Nipkow (1990).

A number of ground convergent systems, including the examples below, have been described by Martin and Nipkow (1990) and Peterson (1990).

Example 4.13. The ground theory of an idempotent, associative, commutative binary operator can be represented by the set of equations

$$
\begin{aligned}
(x \cdot y) \cdot z &\rightarrow x \cdot (y \cdot z) \\
x \cdot y &\approx y \cdot x \\
x \cdot (y \cdot z) &\approx y \cdot (x \cdot z) \\
x \cdot x &\rightarrow x \\
x \cdot (x \cdot y) &\rightarrow x \cdot y
\end{aligned}
$$

which is ground convergent with respect to the lexicographic path ordering.

Example 4.14. *Abelian groups.* The set of equations

$$
\begin{array}{ll}
\begin{aligned}
x + 0 &\rightarrow x \\
0 + x &\rightarrow x \\
x + y &\approx y + x \\
(x + y) + z &\rightarrow x + (y + z) \\
x + (y + z) &\approx y + (x + z)
\end{aligned}
&
\begin{aligned}
x + (-x) &\rightarrow 0 \\
x + (-x + y) &\rightarrow y \\
-(x + y) &\rightarrow -x + -y \\
- - x &\rightarrow x \\
-0 &\rightarrow 0
\end{aligned}
\end{array}
$$

is ground convergent with respect to the lexicographic path ordering $\succ_{lpo}$ corresponding to a precedence $\succ$ in which $- \succ + \succ 0$. (Orientable equations are written as rules.) The reader may want to compare this ground convergent system with the AC-convergent rewrite system given in the preceding chapter.

Example 4.15. *Boolean rings.* The set of equations

$x \oplus 0$	$\rightarrow$	x	$x \wedge 0$	$\rightarrow$	0
$0 \oplus x$	$\rightarrow$	x	$0 \wedge x$	$\rightarrow$	0
			$x \wedge 1$	$\rightarrow$	x
			$1 \wedge x$	$\rightarrow$	x
$x \oplus x$	$\rightarrow$	0	$x \wedge x$	$\rightarrow$	x
$x \oplus y$	$\approx$	$y \oplus x$	$x \wedge y$	$\approx$	$y \wedge x$
$(x \oplus y) \oplus z$	$\rightarrow$	$x \oplus (y \oplus z)$	$(x \wedge y) \wedge z$	$\rightarrow$	$x \wedge (y \wedge z)$
$x \oplus (x \oplus y)$	$\rightarrow$	y	$x \wedge (x \wedge y)$	$\rightarrow$	$x \wedge y$
$x \oplus (y \oplus z)$	$\approx$	$y \oplus (x \oplus z)$	$x \wedge (y \wedge z)$	$\approx$	$y \wedge (x \wedge z)$

$x \wedge (y \oplus z)$	$\rightarrow$	$(x \wedge y) \oplus (x \oplus z)$
$(x \oplus y) \wedge z$	$\rightarrow$	$(x \wedge z) \oplus (y \wedge z)$

is ground convergent with respect to the lexicographic path ordering $\succ_{lpo}$ corresponding to a precedence $\succ$ in which $\wedge \succ \oplus \succ 1 \succ 0$. A rewrite system for Boolean rings which is convergent modulo associativity and commutativity has been described in the preceding chapter.

4.2. Construction of Convergent Rewrite Systems

Ordered completion has been designed to allow the construction of *ground convergent* sets of equations. In most cases of practical interest, ordered completion will actually succeed in constructing a *convergent* rewrite system for a given equational theory, provided such a rewrite system exists at all and the given reduction ordering is completable.

The results in this section apply to a specific version of ordered completion, which we call *unfailing completion*. By $\mathcal{U}^{\succ}$ we denote the standard completion system $\mathcal{C}^{\succ}$ augmented by the following inference rule:

$$\text{DEDUCTION:} \qquad \frac{E\,;R}{E \cup \{s \approx_\top t\}\,;R} \qquad \begin{array}{l} \text{if } s \leftrightarrow_E u \leftrightarrow_E t,\ s \not\succeq u, \\ \text{and } t \not\succeq u \end{array}$$

Inference systems $\mathcal{U}^{\succ}$ are called *unfailing completion systems*. (As before, we write $E\,;R$ to denote a set of equations $E \cup R$, where E consists of equations $s \approx_\top t$ and R of equations $s \approx_\perp t$.)

Unfailing completion applies not just to complete orderings, but also to completable orderings. This greater flexibility in regard to the ordering comes at the expense of a weaker fairness condition which may require computation of more critical pairs.

Definition 4.16. Let $s \approx t$ and $u \approx v$ be two equations with no variables in common, where some non-variable subterm $s|_p$ of s is unifiable with u,

σ being the most general unifier. We say that the superposition of $u \approx v$ on $s \approx t$ at position p determines a *semi-critical pair* $t\sigma \approx s\sigma[v\sigma]_p$ (with respect to the ordering $\succ$) if $t\sigma \not\succ s\sigma$ and $v\sigma \not\succ u\sigma$.

By $SP^{\succ}(E)$ we denote the set of all critical pairs with respect to $\succ$ between equations in $E \cup E^{-1}$. It is evident from the definitions that every ordered critical pair is a semi-critical pair, but not vice versa. In other words, $CP^{\succ}(E) \subseteq SP^{\succ}(E)$.

Definition 4.17. A derivation in the unfailing completion system $\mathcal{U}^{\succ}$ is said to be *fair* if the set of semi-critical pairs $SP^{\succ}(E_\infty \cup R_\infty)$ is a subset of $\bigcup_i (E_i \cup R_i)$.

Proposition 4.18. *Let $>$ be any complete reduction ordering (with respect to E) containing $\succ$. If a derivation from E by the unfailing completion system $\mathcal{U}^{\succ}$ is fair, then the corresponding limit $E_\infty \cup R_\infty$ is ground convergent with respect to $>$.*

Proof. The proof is similar to the proof of Theorem 4.11. Observe that $CP^{>}(E_\infty \cup R_\infty)$ is a subset of $SP^{\succ}(E_\infty \cup R_\infty)$ and that any non-deduction inference by $\mathcal{U}^{\succ}$ is also an inference by $\mathcal{U}^{>}$.   Q.E.D.

Let $\mathcal{F}$ and $\mathcal{V}$ be given sets of function symbols and variables, respectively, and let $\mathcal{K}$ be a set of constants disjoint from $\mathcal{F}$. Any reduction ordering $\succ$ on $T(\mathcal{F}, \mathcal{V})$ can be extended to a reduction ordering $\succ_{\mathcal{K}}$ on $T(\mathcal{F} \cup \mathcal{K}, \mathcal{V})$ as follows.

Let c be any minimal (with respect to $\succ$) constant of $\mathcal{F}$ and let φ be the mapping from $T(\mathcal{F} \cup \mathcal{K}, \mathcal{V})$ to $T(\mathcal{F}, \mathcal{V})$ for which $\varphi(t)$ is the result of replacing in t every constant of $\mathcal{K}$ by c. Moreover, let $\succ_{lpo}$ be the lexicographic path ordering based on some total precedence ordering on $\mathcal{F} \cup \mathcal{K}$. We define, for all terms s and t in $T(\mathcal{F} \cup \mathcal{K}, \mathcal{V})$: $s \succ_{\mathcal{K}} t$ if either $\varphi(s) \succ \varphi(t)$, or else $\varphi(s) = \varphi(t)$ and $s \succ_{lpo} t$.

Lemma 4.19. *Let $\succ$ be a reduction ordering on $T(\mathcal{F}, \mathcal{V})$ and let $\succ_{\mathcal{K}}$ be as described above.*

(1) The binary relation $\succ_{\mathcal{K}}$ is a reduction ordering.

(2) The restriction of $\succ_{\mathcal{K}}$ to terms in $T(\mathcal{F}, \mathcal{V})$ coincides with $\succ$.

(3) Let E be any set of equations, such that $s \leftrightarrow_E t$ implies $\varphi(s) \leftrightarrow_E \varphi(t)$, for all ground terms s and t in $T(\mathcal{F} \cup \mathcal{K})$. Then $\succ_{\mathcal{K}}$ is complete with respect to E if and only if $\succ$ is complete with respect to E.

Proof. (1) The relation $\succ_{\mathcal{K}}$ can easily be shown to be transitive and irreflexive. It is also well-founded. For suppose there is an infinite sequence $t_1 \succ_{\mathcal{K}} t_2 \succ_{\mathcal{K}} \cdots$ of terms in $T(\mathcal{F} \cup \mathcal{K}, \mathcal{V})$. Then there is a corresponding sequence $\varphi(t_1) \succeq \varphi(t_2) \succeq \ldots$ of terms in $\mathcal{T}(\mathcal{F}, \mathcal{V})$. Since $\succ$ is well-founded, we have $\varphi(t_j) = \varphi(t_{j+1}) = \cdots$ for some j. But then there is an infinite sequence $\varphi(t_j) \succ_{lpo} \varphi(t_{j+1}) \succ_{lpo} \cdots$ which contradicts the well-foundedness of the lexicographic path ordering. We conclude that $\succ_\mathcal{K}$ is a well-founded ordering. It is also a rewrite relation.

Let s and t be arbitrary terms for which $s \succ_\mathcal{K} t$. We will show that $u[s\sigma]_p \succ_\mathcal{K} u[t\sigma]_p$, for all terms u, position p in u, and substitutions σ. First note that $s \succ_\mathcal{K} t$ implies $\varphi(s) \succeq \varphi(t)$. Thus

$$\varphi(u[s\sigma]_p) = \varphi(u)[\varphi(s)\sigma']_p \succeq \varphi(u)[\varphi(t)\sigma']_p = \varphi(u[t\sigma]_p),$$

where $x\sigma' = \varphi(x\sigma)$, for all variables x. If $\varphi(s) \succ \varphi(t)$, then obviously $\varphi(u[s\sigma]_p) \succ \varphi(u[t\sigma]_p)$, and therefore $u[s\sigma]_p \succ_\mathcal{K} u[t\sigma]_p$. On the other hand, if $\varphi(s) = \varphi(t)$ then $\varphi(u[s\sigma]_p) = \varphi(u[t\sigma]_p)$. But also $s \succ_{lpo} t$, and hence $u[s\sigma]_p \succ_{lpo} u[t\sigma]_p$, so that again $u[s\sigma]_p \succ_\mathcal{K} u[t\sigma]_p$. We conclude that $\succ_\mathcal{K}$ is a reduction ordering.

(2) Follows immediately from the definition of $\succ_\mathcal{K}$.

(3) Let E be a set of equations, such that $s \leftrightarrow_E t$ implies $\varphi(s) \leftrightarrow_E \varphi(t)$, for all ground terms s and t in $\mathcal{T}(\mathcal{F} \cup \mathcal{K})$.

If $\succ_\mathcal{K}$ is complete with respect to E, then any two distinct ground terms s and t in $\mathcal{T}(\mathcal{F})$, for which $s \leftrightarrow_E^* t$, are comparable with respect to $\succ_\mathcal{K}$. Using (2) we infer that s and t are also comparable with respect to $\succ$.

On the other hand, suppose $\succ$ is complete with respect to E. Let s and t be any two distinct ground terms in $\mathcal{T}(\mathcal{F} \cup \mathcal{K})$ with $s \leftrightarrow_E^* t$. Then $\varphi(s) \leftrightarrow_E^* \varphi(t)$ and, since $\succ$ is complete with respect to E, either $\varphi(s) \succ \varphi(t)$, or $\varphi(t) \succ \varphi(s)$, or $\varphi(t) = \varphi(s)$. Furthermore, either $s \succ_{lpo} t$ or $t \succ_{lpo} s$, as the lexicographic path ordering is complete. Therefore, we have either $s \succ_\mathcal{K} t$ or $t \succ_\mathcal{K} s$, whenever s and t are distinct.   Q.E.D.

Two sets of equations E and E' are said to be *literally similar* if each equation $s \approx t$ in E is literally similar to an equation $u \simeq v$ in E', and vice versa. Recall that if a rewrite system R is reduced, then no right-hand side of a rule in R and no term encompassed by a left-hand side of a rule in R is reducible by R.

Theorem 4.20. *Let R be a reduced convergent system for E and $\succ$ be a completable (with respect to E) reduction ordering containing R. If $E_\infty \cup R_\infty$ is the limit of a fair and simplifying derivation from E by the unfailing completion system $\mathcal{U}^\succ$, then $E_\infty = \emptyset$ and R_∞ is literally similar to R.*

Proof. Let $\mathcal{F}$ and $\mathcal{V}$ be the given sets of function symbols and variables, respectively. In addition, let $\mathcal{K}$ be a set of new (Skolem) constants, which contains a unique constant $\hat{x}$ for each variable x in $\mathcal{V}$ and an additional constant c not associated with any variable. Let $\succ$ be the given reduction ordering. By our assumption, the ordering $\succ$ can be extended to a reduction ordering $>$ on $T(\mathcal{F}, \mathcal{V})$ that is complete with respect to E. The ordering $>$ in turn can be extended to a reduction ordering $>_{\mathcal{K}}$ on $T(\mathcal{F} \cup \mathcal{K}, \mathcal{V})$ that is also complete with respect to E. (Observe that $s \leftrightarrow_E t$ implies $\varphi(s) \leftrightarrow_E \varphi(t)$, for all ground terms s and t in $T(\mathcal{F} \cup \mathcal{K})$.) We may assume, without loss of generality, that the constant c is minimal with respect to $\succ_{\mathcal{K}}$.

By $\hat{t}$ we denote the result of replacing in t all variables by their corresponding Skolem constants. Note that a term t is irreducible by R if and only if its Skolemized version $\hat{t}$ is irreducible by R. Furthermore, as the ordering $>_{\mathcal{K}}$ contains R, a term in $T(\mathcal{F} \cup \mathcal{K}, \mathcal{V})$ is irreducible if and only if it is minimal (with respect to $>_{\mathcal{K}}$) in its congruence class.

Let now $E_\infty ; R_\infty$ be the limit of a fair and simplifying derivation from $E ; \emptyset$. We first show that all left-hand sides of R are reducible by R_∞. If $s \to t$ is a rewrite rule in R, then $\hat{s} \leftrightarrow_{E_\infty} \hat{t}$, so that by Proposition 4.18, there exists a ground rewrite proof of $\hat{s} \approx \hat{t}$ in $E_\infty^{\geq \kappa} \cup R_\infty^{\geq \kappa}$. The term $\hat{t}$ is irreducible by R and therefore minimal in its congruence class. As a consequence, any ground rewrite proof of $\hat{s} \approx \hat{t}$ has to be of the form $\hat{s} \to^*_{E_\infty^{\geq \kappa} \cup R_\infty^{\geq \kappa}} \hat{t}$.

Let P be one such proof that is minimal with respect to the proof ordering induced by $>_{\mathcal{K}}$. If the first proof step of P is by a rule in R_∞, then $\hat{s}$ (and hence s) is reducible by R_∞, and we are done. Let us therefore assume that the first proof step is by an equation in E_∞. Since all proper subterms of $\hat{s}$ are irreducible by R, and therefore minimal in their respective congruence classes, the first proof step in P must be of the form $\hat{s} \leftrightarrow_{u \approx v}^{\lambda} w$. (That is, the equation $u \approx v$ applies at the top.) This implies that $\hat{s}$ and s are instances of u. All proper instances of s are irreducible, and hence minimal in their congruence class. Thus, if s were a proper instance of u, then u would be irreducible by R and $v \succ u$, which contradicts $\hat{s} >_{\mathcal{K}} w$. In short, s and u have to be literally similar.

Since $u \approx v$ is an equation in E_∞ (and the corresponding derivation is simplifying), the two terms u and v are incomparable with respect to $\succ$, so that $u \not\succ v$. Thus P must contain at least two proof steps. Let

$$Q = \hat{s} \leftrightarrow_{u \approx v}^{\lambda} w \leftrightarrow_{u' \approx v'}^{p} w[v'\tau]_p$$

be the first two proof steps in P and and let σ be a substitution, such that $\hat{s} = u\sigma$ and $w = v\sigma$. Since the first proof step applies at the top, the proof Q is a variable or a proper overlap. A variable overlap can be ruled out, as $x\sigma$ is irreducible for all variables x occurring in u or v. (If x occurs in u, then $x\sigma$ is a constant in $\mathcal{K}$, and hence is irreducible; if x occurs in v, but not in u, then $x\sigma = c$ because of the minimality of the proof P.) Suppose Q is a proper overlap.

Let σ' be a substitution, such that for all variables x occurring in v but not in u, $x\sigma' = z$, where z is a variable occurring neither in u nor in v; and $y\sigma' = y$, otherwise. Then $u\sigma' = u$ and there exists a proof

$$u \leftrightarrow^{\lambda}_{u\approx v} v\sigma' \leftrightarrow^{p}_{u'\approx v'} v\sigma'[v'\tau']_p$$

which is a critical overlap with respect to $\succ$ (i.e., $u \not\succeq v\sigma'$ and $v'\tau' \not\succeq u'\tau'$) and hence determines a semi-critical pair $u \approx v\sigma'[v'\tau']_p$. (Here τ' is such that $u'\tau'\rho = u'\tau$ and $v'\tau'\rho = v'\tau$, for some substitution ρ.) Since $(u'\tau)\rho' >_{\mathcal{K}} (v'\tau)\rho'$, for some substitution ρ', we clearly have $v'\tau \not\succeq u'\tau$. On the other hand, if $v\sigma' = v$, then $u \not\succeq v\sigma'$; and if $v\sigma' \neq v$, then v contains some variable not occurring in u, so that again $u \not\succeq v\sigma'$. Thus $u \approx v\sigma'[v'\tau']_p$ is indeed a semi-critical pair of $E_\infty \cup R_\infty$. The proof

$$Q' = \hat{s} \leftrightarrow^{\lambda}_{u\approx v\sigma'[v'\tau']_p} w[v'\tau]_p$$

is less complex than Q, which contradicts that P is a minimal proof of $\hat{s} \approx \hat{t}$. In short, $u \approx v$ cannot be an equation in E_∞, but has to be a rule in R_∞.

We have thus shown that all left-hand sides of R are reducible by R_∞, from which we may easily infer that $E_\infty = \emptyset$ and that R is literally the same as R_∞.   Q.E.D.

The above theorem applies to completable reduction orderings. Reduction orderings induced by reduced convergent rewrite systems need not be completable, however. For example, the rewrite system R, consisting of rules $f(h(x)) \rightarrow f(i(x))$, $g(i(x)) \rightarrow g(h(x))$, $h(a) \rightarrow c$, and $i(a) \rightarrow c$, is reduced and convergent. Any complete reduction ordering $>$ for R must be such that $h(a) > i(a)$ or $i(a) > h(a)$. If $h(a) > i(a)$, then $g(h(a)) > g(i(a))$; while from the second rule in R we infer $g(i(a)) > g(h(a))$. A similar contradiction can be derived from the assumption $i(a) > h(a)$.

Devie (1990) has proved a result similar to the above theorem which requires not that the given ordering be completable, but instead that R be a set of linear equations. We next characterize a class of rewrite systems R for which the rewrite relation $\rightarrow_R$ is completable (with respect to R).

A *reduction sequence* (of length n) in R is any proof of the form $t_0 \rightarrow_R t_1 \rightarrow_R \cdots \rightarrow_R t_n$. If R is finite and terminating, then (by König's Lemma) there are only finitely many reduction sequences with fixed initial term t. A reduction sequence $t_0 \rightarrow^{p_1}_{r_1} t_1 \cdots \rightarrow^{p_{n-1}}_{r_{n-1}} t_{n-1} \rightarrow^{p_n}_{r_n} t_n$ is called *innermost* if, for all i with $1 \leq i \leq n$, each proper subterm of $t_{i-1}|_{p_i}$ is irreducible by R. We denote by $I(t)$ the length of the shortest innermost reduction sequence from t to a normal form t'. If $\succ$ is a complete reduction ordering with respect to R (but does not necessarily extend R), then we define the ordering $\succ_R^i$ by: $s \succ_R^i t$ if $s \leftrightarrow_R^* t$ and either $I(s) > I(t)$ or else $I(s) = I(t)$ and $s \succ t$.

Lemma 4.21. *If R is a reduced convergent system, then the ordering $\succ_R^i$ contains R and whenever $s \succ_R^i t$, then $u[s]_p \succ_R^i u[t]_p$, for all terms s, t, and u, and positions p in u.*

Proof. If R is reduced, then $I(s) = 1$ and $I(t) = 0$, for every rule $s \rightarrow t$ in R. Hence, $\succ_R^i$ contains R. Now suppose that $s \succ_R^i t$, and let s' be the (unique) normal form of s and t in R. Any shortest innermost reduction sequence from $u[s]$ can be rearranged so that s is reduced to s' before any other rewrite steps are applied. In other words, there is a shortest innermost sequence of the form $u[s] \rightarrow_R^* u[s'] \rightarrow_R^* u'$. Since there exists a corresponding innermost sequence $u[t] \rightarrow_R^* u[s'] \rightarrow_R^* u'$, we conclude that $I(u[s]) \geq I(u[t])$, and consequently $u[s] \succ_R^i u[t]$.   Q.E.D.

The ordering $\succ_R^i$ need not be a rewrite relation. For example, if R is $\{f(x) \rightarrow g(x, x, x), a \rightarrow b\}$, then $f(x) \succ_R^i g(x, x, x)$, but $f(a) \not\succ_R^i g(a, a, a)$. But the restriction of $\succ_R^i$ to ground terms is a rewrite relation.

A term s is said to overlap another term t if it can be unified with some term literally similar to a non-variable subterm of t.

Proposition 4.22. *If R is a reduced convergent rewrite system wherein no variable appears more often in a right-hand side than in the corresponding left-hand side and no left-hand side overlaps any right-hand side, then the reduction ordering $\rightarrow_R^+$ is completable.*

Proof. Let $\succ$ be the transitive closure of the union of the reduction ordering $\rightarrow_R^+$ and the restriction of $\succ_R^i$ to ground terms. We claim that this ordering is a well-founded rewrite relation.

To prove well-foundedness, we first show that $s\sigma \succ_R^i t\sigma$, for every ground instance $s\sigma \rightarrow t\sigma$ of a rule in R. Let us denote by σ' the substitution that maps each variable x to the normal form of $x\sigma$. Since no variable appears more often in a right-hand side of a rule than in the corresponding left-hand side, no shortest innermost reduction sequence $t\sigma \rightarrow_R \cdots \rightarrow_R t\sigma'$ can be longer than a shortest innermost sequence $s\sigma \rightarrow_R \cdots \rightarrow_R s\sigma'$. Since no left-hand side of R overlaps any right-hand side, the term $t\sigma'$ is irreducible in R, whereas $s\sigma'$ is reducible. Thus $I(s\sigma) > I(t\sigma)$, which implies $s\sigma \succ_R^i t\sigma$.

In sum, we have proved that $s \rightarrow_R t$ implies $s \succ_R^i t$, for all ground terms s and t. Consequently, the ordering $\succ$ is terminating on ground terms, and hence well-founded.   Q.E.D.

As an immediate corollary we obtain:

Corollary 4.23. (Dershowitz and Marcus 1985) *If a ground rewrite system is convergent and reduced, then it is contained in some complete reduction ordering.*

4.3. Refutational Theorem Proving

Standard completion is primarily a tool for constructing convergent rewrite systems, but has also been used as an equational theorem prover. In this regard the possibility of failure is detrimental, as provability in equational theories is semi-decidable. Ordered completion, on the other hand, is refutationally complete for equational theories, and hence provides a semi-decision procedure for the provability problem.

Let $\mathcal{F}$ be a given set of function symbols, $\mathcal{V}$ be a given set of variables, and E be a set of equations between terms in $T(\mathcal{F}, \mathcal{V})$. Suppose we want to check whether an equation $s \approx t$ is provable in E. Take the Skolemized version $\hat{s} \approx \hat{t}$ of $s \approx t$ (i.e., $\hat{s}$ and $\hat{t}$ are obtained from s and t, respectively, by replacing each variable by a unique Skolem constant) and let $\succ$ be a complete reduction ordering on $T(\mathcal{F} \cup \mathcal{K}, \mathcal{V})$, where $\mathcal{K}$ is the set of Skolem constants occurring in $\hat{s}$ and $\hat{t}$. If E_∞ is the limit of any fair derivation by the ordered completion system $\mathcal{O}^\succ$ from the initial set E, then E_∞ is ground convergent with respect to $\succ$. Now, if $s \approx t$ is provable in E, then $\hat{s} \approx \hat{t}$ is also provable in E and hence there exists a ground rewrite proof of $\hat{s} \approx \hat{t}$ in $E_\infty^\succ$. Conversely, if $\hat{s} \approx \hat{t}$ is provable in E_∞, then $s \approx t$ is provable in E. In short, $s \approx t$ is provable if and only if the two Skolemized terms $\hat{s}$ and $\hat{t}$ can be rewritten to a common normal form by $E_\infty^\succ$. Ordered completion thus provides a semi-decision procedure for the provability problem in equational theories.

Superposition, which forms the basis for critical pair computations, is essentially an ordered version of paramodulation (Robinson and Wos 1969). Ordered completion also uses the given reduction ordering to determine which equations are redundant and can be deleted. In practice, deduction and deletion are combined in various ways, to allow for simplification of equations (via normalization of terms). Systematic simplification may considerably reduce the search space of a proof procedure, without destroying refutation completeness.

The idea of extending standard completion by computing equational consequences of unorientable equations can be traced back to the work of Brown (1975) and Lankford (1975) on integrating resolution and simplification by rewriting (for general first-order clauses with equality). Peterson (1983) proved the refutation completeness of an inference system combining resolution, paramodulation, and simplification with respect to orderings isomorphic to ω on ground terms. (This class of orderings excludes many important orderings, such as most path orderings.) A similar inference system for more general orderings, but without simplification, has been considered by Hsiang and Rusinowitch (1986). Further results in this direction have been obtained by Bachmair and Ganzinger (1990a, b) and Rusinowitch (1991). Fribourg (1985) proved the completeness of a restricted version of paramodulation with locking resolution.

Hsiang and Rusinowitch (1987) used transfinite semantic trees to prove the refutation completeness of an unfailing completion procedure, with a simplification rule that is (due to the structure of transfinite semantic trees) based on the subterm ordering, rather than on the encompassment ordering.[2]

Implementations of unfailing completion have been reported by Mzali (1986) and Ohsuga and Sakai (1986). Anantharaman, Hsiang, and Mzali (1989) have implemented a procedure that combines unfailing completion with associative-commutative completion. Experiments with this prover have been described by Anantharaman and Hsiang (1990).

4.4.   Horn Clauses with Equality

Ordered completion, as a technique for constructing ground convergent sets of equations, can be applied to the more general context of Horn clauses with equality. We will show that the combination of ordered completion with an ordered version of unit resolution is a refutationally complete theorem proving method for such clauses.

Let $\mathcal{F}$ and $\mathcal{V}$ be given sets of function symbols and variables, respectively. In addition, let $\mathcal{P}$ be a set of *predicate symbols*, disjoint from $\mathcal{F}$ and $\mathcal{V}$. We assume that $\mathcal{P}$ contains the binary symbol $\approx$ (which denotes equality).

By an *atomic formula* (or an *atom*) we mean an expression $P(t_1, \ldots, t_n)$, where $t_1, \ldots, t_n$ are terms in $\mathcal{T}(\mathcal{F}, \mathcal{V})$ and P is an n-ary predicate symbol in $\mathcal{P}$. Atomic formulas with the equality predicate are written in infix notation $s \approx t$ and called equality atoms. The set of ground atoms is called the *Herbrand base*; the set of ground terms $\mathcal{T}(\mathcal{F})$, the *Herbrand universe*.

A *clause* is a pair of multisets of literals, usually written $\Gamma \rightarrow \Delta$. The multiset Γ is called the *antecedent*; the multiset Δ, the *succedent*. The letters A and B are used to denote atoms; the letters C and D, to denote

---

[2] Subterm-based simplification is too weak for practical purposes. In particular, simplification of one side of an equation at the *top* is not always possible in the procedure proposed by Hsiang and Rusinowitch. For example, the equation $(x \cdot y) \cdot (y^- \cdot z) \approx x \cdot z$ cannot be simplified by the rule $x \cdot (y \cdot z) \rightarrow (x \cdot y) \cdot z$.

clauses. We usually write Γ_1, Γ_2 instead of $\Gamma_1 \cup \Gamma_2$; Γ, A or A, Γ instead of $\Gamma \cup \{A\}$; and $A_1, \ldots, A_m \rightarrow B_1, \ldots, B_n$ instead of $\{A_1, \ldots, A_m\} \rightarrow \{B_1, \ldots, B_n\}$. A clause $A_1, \ldots, A_m \rightarrow B_1, \ldots, B_n$ represents an implication $A_1 \wedge \cdots \wedge A_m \supset B_1 \vee \cdots \vee B_m$. The empty clause indicates a contradiction. Clauses of the form $\Gamma, A \rightarrow \Delta, A$ are called *tautologies*.

A clause is called a *Horn clause* if its succedent contains at most one atom. A *unit* clause is a clause in which antecedent and succedent combined contain only one atom. We speak of a *positive unit* clause, if that atom occurs in the succedent.

Definition 4.24. An (Herbrand) *interpretation* is a subset of the Herbrand base. An interpretation I is said to *satisfy* a ground clause $\Gamma \rightarrow \Delta$ if either $\Gamma \not\subseteq I$ or else $\Delta \cap I \neq \emptyset$. It *satisfies* a (non-ground) clause $\Gamma \rightarrow \Delta$ if it satisfies all ground instances $\Gamma \sigma \rightarrow \Delta \sigma$.

We also say that a clause C is *true in I*, if I satisfies C; and that C is *false in I*, otherwise. For instance, a tautology is satisfied by any interpretation. Clauses which are satisfied by no interpretation are called *unsatisfiable*. The empty clause, for instance, is unsatisfiable. If I satisfies all clauses of a set N, we say that I is a (Herbrand) *model* of N. A set of clauses is *unsatisfiable* if it has no model. We also write $C_1, \ldots, C_n \models C$ to indicate that C is true in every (Herbrand) model of $\{C_1, \ldots, C_n\}$.

We are mainly interested in *equality interpretations*, that is, interpretations that satisfy all clauses

$$
\begin{aligned}
&\rightarrow& x &\approx x \\
x \approx y &\rightarrow& y &\approx x \\
x \approx y, y \approx z &\rightarrow& x &\approx z \\
x \approx y &\rightarrow& f(\ldots, x, \ldots) &\approx f(\ldots, y, \ldots) \\
x \approx y, P(\ldots, x, \ldots) &\rightarrow& P(\ldots, &y, \ldots)
\end{aligned}
$$

where f ranges over all function symbols and P over all predicate symbols. The set of all these Horn clauses is denoted by EQ. We call I an *equality model* of N if it is a model of $N \cup EQ$.

The fundamental inference rule in clausal theorem proving is *resolution*:

$$
\frac{\Gamma \rightarrow \Delta, A \quad B, \Lambda \rightarrow \Pi}{\Gamma \sigma, \Lambda \sigma \rightarrow \Delta \sigma, \Pi \sigma}
$$
if σ is a most general unifier of A and B.

The conclusion of a resolution inference is called a *resolvent* of the respective premises. Resolution is refutationally complete for general clauses (without equality) when combined with the *factoring* rules

$$\frac{\Gamma, A, B \to \Delta}{\Gamma\sigma, A\sigma \to \Delta\sigma} \quad \text{and} \quad \frac{\Gamma \to \Delta, A, B}{\Gamma\sigma \to \Delta\sigma, A\sigma}$$

where σ is a most general unifier of A and B.

Clauses with equality require additional inference rules, called *paramodulation*:

$$\frac{\Gamma \to \Delta, s \approx t \quad \Lambda, A[u] \to \Pi}{\Gamma\sigma, \Lambda\sigma, A\sigma[t\sigma] \to \Delta\sigma, \Pi\sigma} \quad \text{and} \quad \frac{\Gamma \to \Delta, s \approx t \quad \Lambda \to \Pi, A[u]}{\Gamma\sigma, \Lambda\sigma \to \Delta\sigma, \Pi\sigma, A\sigma[t\sigma]}$$

where σ is a most general unifier of s and u and u is not a variable. We also speak of a paramodulation of the clause $\Gamma \to \Delta, s \approx t$ on the clause $\Lambda, A[u] \to \Pi$ (or $\Lambda \to \Pi, A[u]$).

Various restricted versions of resolution and paramodulation (without factoring) can be shown to be refutationally complete for Horn clauses with equality. We shall present one such inference system that also includes simplification and deletion rules.

Definition 4.25. By a (positive) *unit resolution* we mean a resolution inference in which one of the two premises is a (positive) unit clause. A positive unit resolution

$$\frac{\to A \quad B, \Gamma \to B'}{\Gamma\sigma \to B'\sigma}$$

is said to be *ordered* (with respect to $\succ$) if the atom $B\sigma$ is maximal in $\Gamma\sigma$ (i.e., there is no atom B'' in Γ such that $B''\sigma \succ B\sigma$).

By $PR^{\succ}(N)$ we denote the set of all ordered positive unit resolvents obtainable from clauses in N. Positive unit resolution is a refutationally complete inference rule for Horn clauses without equality (Henschen and Wos 1974).

Definition 4.26. Paramodulation inferences

$$\frac{\to s \approx t \quad A[u], \Gamma \to B}{A\sigma[t\sigma], \Gamma\sigma \to B\sigma} \quad \text{or} \quad \frac{\to s \approx t \quad \to B[u]}{\to B\sigma[t\sigma]}$$

are called *positive unit paramodulations*. Such a paramodulation inference is said to be *ordered* (with respect to $\succ$) if (i) $A\sigma$ is maximal in $\Gamma\sigma$, and (ii) there exists a (ground) substitution τ, such that $(s\sigma)\tau \succ (t\sigma)\tau$ and, whenever $A[u]$ or $B[u]$ is an equality atom $v[u] \approx w$, then $(v\sigma)\tau \succ (w\sigma)\tau$.

By $PP^{\succ}(N, N')$ we denote the set of clauses obtained by ordered paramodulation of a positive unit clause in N on a clause in N'.

In this section, by a *complete* reduction ordering we mean a well-founded ordering $\succ$ on atoms and terms, such that (i) $s \succ t$ implies $u[s\sigma]_p \succ u[t\sigma]_p$ and $P(\ldots, s\sigma, \ldots) \succ P(\ldots, t\sigma, \ldots)$, for all terms s, t, and u, positions p in u, substitutions σ, and predicate symbols P; and (ii) the ordering is total on ground terms (the Herbrand universe) and ground atoms (the Herbrand base).

The inference system $\mathcal{H}^\succ$ (or simply $\mathcal{H}$ if $\succ$ is clear from the context) consists of the inference rules below, which apply to pairs $N \,;E$ where N is a set of Horn clauses and E a set of equations. If E is a set of equations, we also denote by $C(E)$ the set of all positive unit clauses $\to s \approx t$, such that $s \approx t \in E$.

(1) Inference rules applied to equations:

DEDUCTION:
$$\frac{N\,;E}{N\,;E \cup \{s \approx_n t\}} \qquad \text{if } s \leftrightarrow^*_E t$$

DELETION:
$$\frac{N\,;E \cup \{s \approx_n t\}}{N\,;E} \qquad \begin{array}{l}\text{if there is a proof } P \text{ of}\\ s \approx t \text{ in } E \text{ such that}\\ s \leftrightarrow^\lambda_{s \approx_n t} t \succ_O P\end{array}$$

(2) Inference rules to convert positive unit clauses to equations:

CONVERSION:
$$\frac{N \cup \{\to s \approx t\}\,;E}{N\,;E \cup \{s \approx t\}}$$

(3) Inference rules applied to Horn clauses:

RESOLUTION:
$$\frac{N\,;E}{N \cup \{C\}\,;E} \qquad \begin{array}{l}\text{if } C \text{ is a resolvent of}\\ \text{clauses in } N \cup \{\to x \approx x\}\end{array}$$

PARAMODULATION:
$$\frac{N\,;E}{N \cup \{C\}\,;E} \qquad \begin{array}{l}\text{if } C \text{ is a paramodulant}\\ \text{of a clause in } C(E) \text{ on a}\\ \text{clause in } N\end{array}$$

SUBSUMPTION:
$$\frac{N \cup \{C, D\}\,;E}{N \cup \{C\}\,;E} \qquad \begin{array}{l}\text{if } C \text{ properly subsumes}\\ D\end{array}$$

DELETION:
$$\frac{N \cup \{C\}\,;E}{N\,;E} \qquad \begin{array}{l}\text{if } C \text{ is redundant in } N \cup\\ C(E)\end{array}$$

A clause $C = \Gamma \to \Delta$ is said to *subsume* a clause D if D is of the form $\Gamma\sigma, \Lambda \to \Delta\sigma, \Pi$. If, in addition, D does not subsume C, we say that C *properly subsumes* D.

The concept of redundancy of clauses is adapted from (Bachmair and Ganzinger 1990a). Let $\succ^s$ be an ordering on Horn clauses defined by: $(\Gamma \to A) \succ^s (\Delta \to B)$ if either $\Gamma \succ_{mul} \Delta$ or else $\Gamma = \Delta$ and $A \succ B$.

Definition 4.27. A clause C is said to be *redundant* in a set of clauses N (with respect to the reduction ordering $\succ$) if for every ground instance $C\sigma$ of C, there exist ground instances $C_1, \ldots, C_n$ of clauses in N, such that (i) $C_1, \ldots, C_n \models C\sigma$ and (ii) $C\sigma \succ^s C_i$, for all i.

Tautologies, for instance, are trivially redundant. More complicated simplification techniques can be described by inference rules derived from deletion, resolution, and paramodulation. For further details we refer to Bachmair and Ganzinger (1990b).

For example, the inference rule

SIMPLIFICATION: $$\frac{N \cup \{C[s\sigma]\}; E \cup \{s \approx t\}}{N \cup \{C[t\sigma]\}; E \cup \{s \approx t\}} \quad \text{if } C[s\sigma] \succ C[t\sigma]$$

is obtained by combining paramodulation and deletion.

Inference system $\mathcal{H}^\succ$ are sound in the following sense.

Lemma 4.28. *If* $N ; E \vdash_{\mathcal{H}} N' ; E'$, *then every Herbrand model of* $N \cup C(E) \cup EQ$ *is also a model of* $N' \cup C(E') \cup EQ$.

Proof. Resolvents and paramodulants are logical consequences of their respective premises. Also, if $s \approx t$ is true in an equality interpretation I, then $u[s\sigma]_p \approx u[t\sigma]_p$ is true in I.   Q.E.D.

Definition 4.29. A derivation in $\mathcal{H}^\succ$ is said to be *fair* if
(i) the set of ordered critical pairs $CP^\succ(E_\infty)$ is a subset of the set of all derived equations $\bigcup_i E_i$;
(ii) the set of ordered positive unit resolvents $PR^\succ(N_\infty \cup \{\to x \approx x\})$ is a subset of the set of all derived Horn clauses $\bigcup_i N_i$;
(iii) the set of ordered paramodulants $PP^\succ(C(E_\infty), N_\infty)$ is a subset of the set of all derived Horn clauses $\bigcup_i N_i$; and
(iv) N_∞ contains no positive unit clause of the form $\to s \approx t$.

Fair derivations by ordered completion component of $\mathcal{H}$ result in ground convergent sets of equations.

Lemma 4.30. *Let* $\succ$ *be a complete reduction ordering. If* $N_\infty \cup E_\infty$ *is the limit of a fair derivation in* $\mathcal{H}^\succ$, *then* E_∞ *is ground convergent with respect to* $\succ$.

In addition, we will prove that $\mathcal{H}$ is refutationally complete for Horn clauses with equality, for which purpose we need the following (well-known) lemma.

Lemma 4.31. (Lifting Lemma) *Let $C\sigma$ and $D\sigma$ be ground instances of clauses C and D, respectively.*

(1) *Every (ordered positive unit) resolvent of $C\sigma$ and $D\sigma$ is a ground instance of some (ordered positive unit) resolvent of C and D.*

(2) *Let C be a clause $\rightarrow s \approx t$ and D be a clause $\Lambda, A[u] \rightarrow \Pi$, such that u is not a variable (and C and D have no variables in common). Suppose $s\sigma = u\sigma$ and C' is a (ordered) paramodulant $\Gamma\sigma, A\sigma[t\sigma] \rightarrow \Delta\sigma$ of $C\sigma$ on $D\sigma$. Then C' is a ground instance of a (ordered) paramodulant of C on D.*

A proof of the lemma can be found in Peterson (1983).

Theorem 4.32. (Refutation completeness) *Let $\succ$ be a complete reduction ordering and N be a set of Horn clauses. If $N_\infty \cup E_\infty$ is the limit of a fair derivation in $\mathcal{H}^\succ$ from N, then $N \cup EQ$ is unsatisfiable if N_∞ contains the empty clause.*

Proof. By soundness, every model of N is also a model of N_∞. Thus, if N_∞ contains the empty clause, then $N \cup EQ$ has no model.

Suppose, on the other hand, that N_∞ does not contain the empty clause. Let I be the Herbrand interpretation for which $P(t_1, \ldots, t_n) \in I$ if and only if there exists a ground instance $\rightarrow P(s_1, \ldots, s_n)$ of some unit clause in $N_\infty \cup \{\rightarrow x \approx x\}$, such that $t_i \rightarrow^*_{E_\infty^\succ} s_i$, for all i. (By this definition, a ground equality atom $s \approx t$ is in I if and only if there exists a ground rewrite proof of $s \approx t$ in $E_\infty^\succ$.) We claim that I is a model of $EQ \cup \bigcup_i (N_i \cup C(E_i))$.

It can easily be seen from the definition that I is an equality interpretation, and therefore is a model of EQ. Furthermore, any ground instance of an equation in E_∞ is contained in I, which implies that I is a model of every set of unit clauses $C(E_i)$. In addition, all unit clauses of N_∞ are satisfied by I. It remains to be proved that I is a model of every set of Horn clauses N_i.

Suppose some ground instance of a clause in $\bigcup_i N_i$ is not satisfied by I. Let C be a minimal such ground instance with respect to the ordering $\succ^s$. Furthermore, let C' be a clause in $\bigcup_i N_i$ and σ be a substitution, such that $C = C'\sigma$. We may assume, without loss of generality, that C' is not properly subsumed by any other clause in $\bigcup_i N_i$. Also note that $x\sigma$ is irreducible by $E_\infty^\succ$, for all variables x, as otherwise C would not be a minimal clause false in I. In other words, the substitution σ is irreducible. This will allow us to apply the Lifting Lemma in all cases below.

Now suppose C' is not a clause in N_∞. Since C' is not properly subsumed by any other clause, this means that C' was either converted to an equation or was deleted at some point in the derivation. If C' was converted to an equation, then C cannot be false in I. On the other hand, if C' was deleted at some point, then there exist clauses $C_1, \ldots, C_n$, such that $C_1, \ldots, C_n \models C$ and $C \succ^s C_i$, for all i. Since C is false in I, some ground clause C_i also has to be false in I, contradicting the minimality of C. Thus C' has to be a clause in N_∞.

Consequently, C is some (non-unit) clause $A_1, \ldots, A_k \to A$, where $k \geq 1$, $A \notin I$, and $\{A_1, \ldots, A_k\} \subseteq I$. Let us assume, without loss of generality, that $A_1 \succeq A_2 \succeq \cdots \succeq A_k$.

If A_1 is an equality atom of the form $t \approx t$, then there is an inference

$$\frac{\to t \approx t \quad t \approx t, A_2, \ldots, A_k \to A}{A_2, \ldots, A_k \to A}$$

by ordered resolution. By the fairness assumption and the Lifting Lemma, the conclusion D of this inference is a ground instance of some clause in $\bigcup_i N_i$. But D is false in I and $C \succ^s D$, which contradicts our assumption that C is a minimal false clause.

Suppose A_1 is an equality atom of the form $s \approx t$, where $s \succ t$. Since $s \approx t$ is true in I, there exists a ground rewrite proof of $s \approx t$ in $E_\infty^\succ$. The term s is thus reducible by $E_\infty^\succ$; and hence there is an inference

$$\frac{\to u \approx v \quad s[u]_p \approx t, A_2, \ldots, A_k \to A}{s[v]_p \approx t, A_2, \ldots, A_k \to A}$$

by ordered paramodulation. Using the Lifting Lemma and the fairness assumption, we may again infer that the conclusion D of this inference is a ground instance of some clause in $\bigcup_i N_i$, which contradicts that C is a minimal false clause.

Finally, let us suppose A_1 is a non-equality atom $P(t_1, \ldots, t_n)$. Then there is some unit clause $\to P(s_1, \ldots, s_n)$ in N_∞, such that $t_i \to^*_{E_\infty} s_i$, for all i with $1 \leq i \leq n$.

If $s_i = t_i$, for all i, then there is an ordered resolution inference

$$\frac{\to P(t_i, \ldots, t_n) \quad P(t_1, \ldots, t_n), A_2, \ldots, A_k \to A}{A_2, \ldots, A_k \to A}.$$

On the other hand, if $t_i \to^+_{E_\infty} s_i$, for some i, then t_i is reducible by $E_\infty^\succ$, so that there is an ordered paramodulation inference

$$\frac{\to u \approx v \quad P(t_1, \ldots, t_i[u]_p, \ldots, t_n), A_2, \ldots, A_k \to A}{P(t_1, \ldots, t_i[v]_p, \ldots, t_n), A_2, \ldots, A_k \to A}.$$

In either case we may use the Lifting Lemma and fairness to infer that the conclusion D of the inference is a clause in $\bigcup_i N_i$, which contradicts that C is a minimal false clause.

We conclude that no ground instance of a clause in $\bigcup_i N_i$ is false in I. In particular, I is a model of $N = N_0$.   Q.E.D.

Dershowitz (1990) has used proof orderings to establish a proof normalization result of an inference system for conditional equations that corresponds to a specific version of the system $\mathcal{H}$ above. (The deduction rules are essentially the same as in $\mathcal{H}$, but the simplification rules are weaker and can be derived from deletion, resolution, and paramodulation.)

Paul (1986) has studied the application of standard completion to Horn clauses with equality. The word problem in Horn clause theories has been studied by Kounalis and Rusinowitch (1987).

Summary

We have presented an extension of standard completion, called ordered completion, which is designed to construct a set of equations that is ground convergent with respect to a complete reduction ordering. A fair ordered completion procedure is guaranteed to succeed in constructing a ground convergent system. Furthermore, we have proved that a specific version of ordered completion, called unfailing completion (which also extends standard completion), always succeeds in constructing a convergent (and not only a ground convergent) rewrite system, provided the given reduction ordering is completable.

Ordered completion is also a refutationally complete theorem prover for purely equational theories. We have discussed further applications of the method to Horn clauses with equality, and have obtained a refutationally complete inference system that combines ordered paramodulation (for the equational part) with ordered positive unit resolution and ordered positive unit paramodulation (for the Horn clause part) and also includes powerful rules for deletion of redundant clauses.

The techniques underlying ordered completion have also been applied to A-unification and "rigid unification," a form of unification of importance in theorem provers based on equational matings; see Gallier and Snyder (1989), Gallier, Snyder, and Raatz (1989) and Gallier, et al. (1988). For another interesting application, to unification in Boolean rings and Abelian groups, see Boudet, Jouannaud, and Schmidt-Schauß (1988). Reddy (1989) found the approach to be of advantage in describing rewrite techniques for program synthesis. Applications to first-order theorem proving (with or without equality) are described in Bachmair and Dershowitz (1987).

5. PROOF BY CONSISTENCY

In many applications, such as algebraic data type specifications and equational programming, equations are intended to define a certain standard model, called the "initial model." Reasoning about algebraic data types and equational programs thus requires proof methods for this initial algebra semantics. Such proof methods typically employ some induction scheme, e.g., induction on the structure of terms. We shall discuss an alternative approach—proof by consistency—that can be applied to equational theories that are presented as ground convergent rewrite systems.

5.1. Consistency and Ground Reducibility

Definition 5.1. An equation $s \approx t$ is said to be an *inductive theorem* of a set of equations E, written $s =_{\mathcal{I}(E)} t$, if $s\sigma \leftrightarrow^*_E t\sigma$, for all ground equations $s\sigma \approx t\sigma$.

For example, associativity and commutativity, $(x + y) + z \approx x + (y + z)$ and $x + y \approx y + x$, are inductive theorems of $R_0 = \{x + 0 \to x, x + S(y) \to S(x + y)\}$. (They are *not* theorems, though, as there are models of R_0 in which the function denoted by $+$ is not associative and commutative.)

Dershowitz (1982a) has pointed out that an equation $s \approx t$ is an inductive theorem of a (ground) convergent rewrite system R if and only if no equation $u \approx v$, where u and v are distinct ground terms irreducible by R, is provable in $R \cup \{s \approx t\}$. Thus, if $s \approx t$ is an inductive theorem of R, then any (ground) convergent rewrite system for $R \cup \{s \approx t\}$ defines the same ground normal forms as R.[1]

Let R be a terminating rewrite system that is ground convergent with respect to the reduction ordering $\to^+_R$. (That is, for all ground terms s and t with $s \leftrightarrow^*_R t$, there exists a ground term v, such that $s \to^*_R v \leftarrow^*_R t$.) We also write $s \to_{R!} t$ if $s \to^*_R t$ and t is in normal form.

Definition 5.2. We say that an equation $s \approx t$ is *consistent with* R if, for every ground equation $s\sigma \approx t\sigma$, the two terms $s\sigma$ and $t\sigma$ can be reduced

---

[1] If R is ground convergent, then the algebra defined on the set of ground normal-form terms is an initial model of R (Goguen 1980).

99

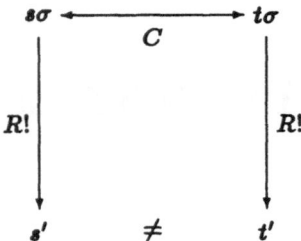

Figure 5.1: *Inconsistency*

to identical normal forms by R. Otherwise, $s \approx t$ is said to be *inconsistent* with R.

A set of equations C is said to be consistent with R if all its equations are consistent with R; and inconsistent, otherwise (see Figure 5.1). Roughly speaking, consistency expresses in proof-theoretical terms that the two sets of equations R and $R \cup \{s \approx t\}$ define the same initial algebra.

Proposition 5.3. (Dershowitz 1982a) *A set of equations C is consistent with a ground convergent rewrite system R if and only if all equations in C are inductive theorems of R.*

Proof. Let $s \approx t$ be any equation in C. If C is consistent with R, then $s\sigma \leftrightarrow^*_R t\sigma$, for all ground equations $s\sigma \approx t\sigma$. Consequently, $s \approx t$ is an inductive theorem of R.

On the other hand, suppose that all equations in C are inductive theorems of R. If $s\sigma \approx t\sigma$ is a ground instance of an equation in C, then $s\sigma \leftrightarrow^*_R t\sigma$ and, since R is ground convergent, the two terms $s\sigma$ and $t\sigma$ can be reduced to a common normal form by R, which implies that C is consistent with R.   Q.E.D.

Consistency—in the above sense—is not decidable. However, we will show that any (finite) inconsistent set of equations C can be transformed into a "provably inconsistent" set. The key here is the notion of "ground reducibility."

Definition 5.4. (Jouannaud and Kounalis 1986). *A term t is ground reducible by R if all its ground instances are reducible by R.*

For example, if $R_0 = \{x + 0 \to x, x + S(y) \to S(x+y)\}$, then any term in which the function symbol $+$ appears is ground reducible by R_0.

Ground reducibility is decidable for finite rewrite systems (Plaisted 1985; Kapur, Narendran, and Zhang 1987), but is provably in exponential time, even for left-linear rewrite system (Kapur, et. al. 1987). Algorithms for deciding ground-reducibility with respect to left-linear rewrite systems have been described by Jouannaud and Kounalis (1986) and Kapur, Narendran, and Zhang (1986). In theories with free constructors ground reducibility is trivially decidable: a term is ground reducible if and only if it contains a non-constructor symbol (cf. the discussion of constructors in Huet and Hullot 1982; Jouannaud and Kounalis 1986; and Fribourg 1989).

What we actually need, however, is a suitable notion of ground reducibility for equations.

Definition 5.5. We say that an equation $s \approx t$ is *ground reducible* by R if, for every ground equation $s\sigma \approx t\sigma$, the two terms $s\sigma$ and $t\sigma$ are identical whenever they are both irreducible by R.[2]

In other words, an equation $s \approx t$ is *not* ground reducible if there exists a ground equation $s\sigma \approx t\sigma$ for which $s\sigma$ and $t\sigma$ are irreducible, yet distinct.

For example, the equation $x + y \approx y + x$ is ground reducible by R_0, but $S(S(x)) \approx S(x)$ is not, as there is a ground instance, $S(S(0)) \approx S(0)$, in which both terms are irreducible, yet distinct.

Decision procedures for testing ground reducibility of terms can easily be adapted to the problem of checking ground reducibility of equations.

Lemma 5.6. *If C is consistent with R, then all equations in C are ground reducible.*

Proof. Suppose some equation $s \approx t$ in C is not ground reducible by R. Then there has to be some ground instance $s\sigma \approx t\sigma$ wherein both $s\sigma$ and $t\sigma$ are irreducible, yet distinct. Hence C is not consistent with R.   Q.E.D.

The reverse direction is not true: a set of ground reducible equations need not be consistent. For instance, the equation $S(S(x + 0)) \approx S(x + 0)$ is ground reducible by R_0, but not consistent with R_0.

5.2.   Proof by Consistency

We say that two reduction orderings $\succ_1$ and $\succ_2$ are *compatible* if the transitive closure of their union is also a reduction ordering. (In other words, compatibility means that the transitive closure of $\succ_1 \cup \succ_2$ is well-founded.) Let R be a finite ground convergent rewrite system and $\succ$ be a reduction ordering compatible with the reduction ordering $\rightarrow_R^+$.

---

[2] This notion of ground reducibility of equations is different from the notion of "co-reducibility" introduced by Jouannaud and Kounalis (1986).

Definition 5.7. A set of equations C is said to be *provably inconsistent with R* (relative to $\succ$) if it contains either an equation that is not ground reducible by R, or an equation $s \simeq t$, where $s \succ t$ and s is not ground reducible by R.

In other words, if C is provably inconsistent with R, then there exists a proof

$$s' = s\sigma \leftrightarrow^{\lambda}_{s \approx t} t\sigma = t'$$

or

$$s' = s\sigma \leftrightarrow^{\lambda}_{s \approx t} t\sigma \rightarrow_{R!} t'$$

in $R \cup C$, where s' and t' are irreducible by R and $s' \neq t'$. (The inconsistency test used by Jouannaud and Kounalis (1986) in their inductive completion procedure is based on the existence of proofs of the latter form.)

Lemma 5.8. *Every provably inconsistent set of equations is inconsistent.*

Proof. Suppose C is provably inconsistent with R. By Lemma 5.6, if some equation in C is not ground reducible by R, then C is inconsistent with R. On the other hand, if C contains an equation $s \simeq t$, such that $s \succ t$ and s is not ground reducible, then there exists a ground proof $s\sigma \leftrightarrow_C t\sigma \rightarrow_{R!} t'$, wherein $s\sigma$ and t' are irreducible. Since $s\sigma \succ t\sigma \succ t'$, the two normal forms $s\sigma$ and t' are distinct, which implies that C is inconsistent. Q.E.D.

It is decidable whether a given (finite) set of equations C is provably inconsistent (provided the ordering $\succ$ is decidable). On the other hand, if C is inconsistent, but not provably inconsistent, then there exists a ground proof P of the form

$$s' \leftarrow_{R!} w \leftarrow_R s\sigma \leftrightarrow^{\lambda}_{s \approx t} t\sigma \rightarrow_{R!} t'$$

where $s \simeq t$ is an equation in C, such that $t \not\succ s$ and s' and t' are distinct irreducible terms. We can deduce new equations from R and C to repeatedly simplify this proof until eventually a provably inconsistent set C' is obtained.

First note that we may assume without loss of generality that the substitution σ is *irreducible*, in the sense that $x\sigma$ is irreducible, for all variables x. For suppose $x\sigma$ were reducible by R, for some variable x. Let ρ be the substitution for which $x\rho$ is the normal form of $x\sigma$, and let P' be the proof

$$s' \leftarrow_{R!} s\rho \leftrightarrow_{s \approx t} t\rho \rightarrow_{R!} t'.$$

We have $s\sigma \rightarrow^+_R s'$ or $s' = s\sigma$. If $s' = s\sigma$, then $s \not\succ t$ and $t\rho \rightarrow^+_R t'$, as otherwise C would be provably inconsistent. In either case, P' is a proof of the desired form.

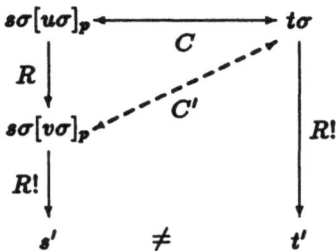

Figure 5.2: *Proof simplification*

Let us therefore assume that P is a proof as above, where in addition σ is an irreducible substitution. Then the subproof

$$w = s\sigma[v\sigma]_p \ {}_{v\approx u}^{p}\!\!\leftarrow\ s\sigma \ \leftrightarrow_{s\approx t}^{\lambda}\ t\sigma$$

is a proper overlap. (We assume, without loss of generality, that $u \approx v$ and $s \approx t$ have no variables in common.) By the Critical Pair Lemma, there exist a critical overlap $s\tau[v\tau]_p \ {}_{v\approx u}^{p}\!\!\leftarrow\ s\tau \ \leftrightarrow_{s\approx t}^{\lambda}\ t\tau$ and a substitution σ', such that $(x\tau)\sigma' = x\sigma$, for all variables x in $s \approx t$ or $u \approx v$. Furthermore, the proof

$$s' \leftarrow_{R!} s\sigma[v\sigma]_p \ \leftrightarrow_{s\tau[v\tau]\approx t\tau}^{\lambda}\ s\sigma \rightarrow_{R!} t'$$

is simpler (according to some well-founded complexity measure on proofs to be defined below) than the proof P (cf. Figure 5.2). This proof simplification process constitutes the basis of our proposed inductive theorem proving method.

Let R be a ground convergent rewrite system and $\succ$ be a reduction ordering compatible with $\rightarrow_R^+$. By $CP(R, C)$ we denote the set of all critical pairs of an equation in R on an equation in C. The inference system $\mathcal{P}_R^\succ$ (or simply $\mathcal{P}$ if R and $\succ$ are clear from the context) consists of the following inference rules, where C may be any set of equations ("conjectures") and

L any set of inductive consequences ("lemmas") of R:

DEDUCTION:
$$\frac{L;C}{L;C\cup\{s\approx t\}} \qquad \text{if } s\approx t\in CP(R,C\cup C^{-1})$$

DELETION:
$$\frac{L;C\cup\{s\approx t\}}{L;C} \qquad \text{if } s =_{\mathcal{I}(R)} t$$

INDUCTION:
$$\frac{L;C}{L\cup\{s\approx t\};C} \qquad \text{if } s =_{\mathcal{I}(R)} t$$

SIMPLIFICATION:
$$\frac{L;C\cup\{s\simeq t\}}{L;E\cup\{u\simeq t\}} \qquad$$
if $s\succ u$ and either $s\leftrightarrow^{+}_{R\cup L}$ u or else $s\leftrightarrow^{p}_{v\approx w}\ u$, for some equation $v\approx w$ in C, such that $s\rhd v$ and $v\succ w$

where the symbol $\rhd$ denotes the strict part of the encompassment ordering. (As usual, we assume that the label of an equation indicates to which of the sets R, L, or C it belongs.)

The deduction rule suffices to derive a provably inconsistent set from any inconsistent set of conjectures. The remaining inference rules are essential, however, for possibly verifying that a given set of conjectures is consistent.

Once a conjecture has been established as an inductive theorem, it can be deleted and then added as a lemma, by the corresponding inference rules. The induction rule can be applied to any inductive theorem $s\approx t$ (even if that theorem was proved by some other proof method). In this sense, proof by consistency can be combined with other inductive theorem proving methods. The inductive theorems in L can be used to simplify a conjecture, but no new equations need to be deduced from them. The introduction of additional lemmas therefore does not increase the search space of proof by consistency.

Each equation can be simplified before any consequences are deduced from it. Simplification by equations in C is slightly more restrictive than necessary. For instance, a conjecture $s\approx t$ in C cannot be simplified by any other conjecture $s\approx u$. (The inference rule can easily be generalized by taking the labels into account, in a similar way as for ordered completion.)

Simplification by equations in $R\cup L$ is unrestricted, and only the final term u, but none of the intermediate terms in the proof $s\leftrightarrow^{+}_{R\cup L} u$ is required to be smaller than s. For instance, if commutativity and associativity are known to be inductive theorems, then equations can be simplified by associative-commutative rewriting. (But in contrast with the associative-commutative inductive completion procedure suggested by Jouannaud and

Kounalis (1986), proof by consistency does not use associative-commutative unification when deducing new equations.)

Inference systems $\mathcal{P}$ are sound in the following sense.

Proposition 5.9. (Soundness) *Let R be a ground convergent system. If $L\,;C \vdash_{\mathcal{P}_R} L'\,;C'$, then C is consistent with R if and only if C' is.*

For example, from the rewrite system $R_0 = \{x + 0 \to x, x + S(y) \to S(x + y)\}$ and the conjecture $x + y \approx y + x$, we can derive new equations $0 + x \approx x$ and $S(x) + y \approx S(x + y)$. The set of all derived equations can be shown to be consistent (as we shall describe below). By soundness, all equations are inductive theorems of R.

Definition 5.10. By a *proof by consistency procedure* we mean any program that accepts as input a ground convergent system R, a reduction ordering $\succ$ compatible with $\to_R^+$, a set L of inductive theorems of R, and a set of conjectures C; and uses the inference rules of $\mathcal{P}_R^\succ$ to generate a derivation from $L \cup C$.

A characteristic of proof by consistency procedures is that they are *linear*, in that all new equations are deduced by superposing rules from the initial rewrite system R on (initial or deduced) conjectures. The procedure by Fribourg (1989) is a proof by consistency procedure in this sense, while most so-called *inductive completion procedures* are not. The latter procedures are essentially standard completion procedures augmented by a consistency check and compute *all* critical pairs in $CP(R \cup C)$, e.g., also critical pairs between conjectures.[3]

Musser (1980) was the first to describe an inductive completion procedure. His procedure applies to abstract data type specifications, where an equality predicate eq_D is associated with each data type D and the specification is "sufficiently complete," so that each ground expression $eq_D(s,t)$ can be reduced to the Boolean constant *true* or *false*. The equation *true* $\approx$ *false* indicates an inconsistency. Various improvements of the basic scheme have been described by Huet and Hullot (1982), Lankford (1981), Dershowitz (1982a), Jouannaud and Kounalis (1986), Fribourg (1989), and Kapur, Narendran, and Zhang (1986). The various approaches mainly differ in the respective notions of consistency they employ.

Huet and Hullot (1982) studied the case of theories with (free) constructors, in which case an inconsistency is signified by an equation between two

---

[3] Given a ground convergent rewrite system R, an equation $s \approx t$ is an inductive theorem of R if the completion procedure constructs a convergent system for $R \cup \{s \approx t\}$ without deriving an inconsistency, while the equation is not an inductive theorem if completion does derive an inconsistency.

distinct ground terms built solely from constructor symbols. In contrast with Musser's method, an explicit axiomatization of equality is not required in this context.

Jouannaud and Kounalis (1986) design an inductive completion procedure based on ground reducibility of terms. An inconsistency is signified by a rewrite rule $u \to v$ the left-hand side of which is not ground reducible. In a similar procedure by Kapur, Narendran, and Zhang (1986) "test sets" are used to check for consistency.

The method proposed by Fribourg (1989) can be viewed as a linear restriction of inductive completion or as a proof by consistency procedure. The use of a linear deduction strategy may considerably restrict the number of equations to be deduced, so that the search space is smaller. In fact, many of the equations that have to be deduced in order to construct a (ground) convergent rewrite system may be irrelevant for proving a given inductive theorem. In essence, full inductive completion corresponds to attempting to solve *all* possible induction schemes—and fails to terminate if one induction schema diverges. Linear proof by consistency procedures, on the other hand, may select one specific induction schema. As a consequence, a proof by consistency procedure may succeed in proving inductive theorems for which inductive completion does not terminate; see Fribourg (1989) for an example. On the other hand, it is also possible that full completion may deduce useful lemmas that cannot be obtained by the more restrictive linear deduction mechanism of a proof by consistency procedure.

Example 5.11. (Küchlin 1989) Let R be the ground convergent rewrite system

$$
\begin{aligned}
app(nil, x) &\to x \\
app(cons(x, y), z) &\to cons(x, app(y, z)) \\
rev(nil) &\to nil \\
rev(cons(x, y)) &\to app(rev(y), cons(x, nil))
\end{aligned}
$$

and suppose $rev(rev(x)) \approx x$ is to be established as an inductive theorem of R.

The last rule in R can be superposed on the given conjecture, resulting in a critical pair

$$rev(app(rev(y), cons(x, nil))) \approx cons(x, y).$$

If we superpose the initial conjecture $rev(rev(x)) \approx x$ on this new conjecture, we obtain a critical pair

$$rev(app(y, cons(x, nil))) \approx cons(x, rev(y))$$

which can be oriented into a rule

$$rev(app(y, cons(x, nil))) \to cons(x, rev(y))$$

(with respect to a suitable lexicographic path ordering). The first critical pair can now be simplified and deleted. The remaining set of rewrite rules is ground convergent, which implies that $rev(rev(x)) \approx x$ is an inductive theorem of R.

On the other hand, with a linear deduction strategy the above rewrite rule cannot be deduced and an infinite derivation

$$rev(app(rev(y), cons(x, nil))) \quad \approx \quad cons(x, y)$$
$$rev(app(app(rev(z), cons(y, nil)), cons(x, nil))) \quad \approx \quad cons(x, cons(y, z))$$
$$\ldots$$

may be produced.

In general, the inference rules of $\mathcal{P}$ do not change the initial ground convergent system R, whereas with inductive completion the goal is the construction of a (ground) convergent rewrite system for $R \cup C$ and the initial system R need not be retained.

One of the main problems with full inductive completion is that, like standard completion, it must be supplied with an ordering on terms which is used to orient equations into rewrite rules and may fail if an equation is generated that cannot be oriented in the given ordering. An important advantage of proof by consistency is that unorientable equations can be handled and any ordering on terms that is compatible with the rewrite relation $\rightarrow_R^+$ may be used. Proof by consistency does not fail and in fact is a complete method for disproving inductive theorems.[4]

5.3.   Refutation Completeness

A proof by consistency procedure is said to be *refutationally complete* if from every inconsistent set of conjectures C_0 it generates a derivation for which some set C_i is provably inconsistent.

The derivation of a provably inconsistent set of equations can be viewed as a proof normalization process. In this context, we are interested in proof transformations that apply to proofs of the form

$$s' \leftarrow_R^* s\sigma \leftrightarrow_C t\sigma \rightarrow_R^* t'$$

where s' and t' are distinct terms irreducible by R. A normal-form proof is a proof $u\tau \leftrightarrow_C v\tau$, wherein $u\tau$ is irreducible and either $v\tau$ is also irreducible and distinct from $u\tau$, or else $u \succ v$. The existence of such a normal-form proof implies that the set C is provably inconsistent.

---

[4]The inference system $\mathcal{P}$ thus solves problem P146 posed by J.-P. Jouannaud in *Bulletin of the EATCS 31*, February 1987.

We first introduce a suitable complexity measure on proofs. The cost of a single proof step $s\sigma \leftrightarrow^{\lambda}_{s\approx t} t\sigma$ is defined to be:

$$\begin{cases} (\{s\sigma\}, s, t\sigma) & \text{if } s \succ t \\ (\{t\sigma\}, t, s\sigma) & \text{if } t \succ s \\ (\{s\sigma, t\sigma\}, \bot, \bot) & \text{otherwise} \end{cases}$$

(The cost assigned to other proof steps, which are not of this form, is irrelevant.)

If P is a ground proof of the form $u \leftarrow^{*}_{R} s\sigma \leftrightarrow^{\lambda}_{s\approx t} t\sigma \rightarrow^{*}_{R} v$, where u and v are not equivalent in R, then the complexity γ^{P} of P is the cost of the proof step $s\sigma \leftrightarrow^{\lambda}_{s\approx t} t\sigma$. If P is the empty proof, then $\gamma^{P} = (\top, \bot, \bot)$. In all other cases, $\gamma^{P} = (\top, \top, \top)$.

If $\succ$ is a reduction ordering compatible with $\rightarrow^{+}_{R}$, we denote by $\succ_{R}$ the transitive closure of the union of $\succ$ and $\rightarrow^{+}_{R}$; and by $\rhd_{R}$ the transitive closure of the union of $\succ_{R}$ with the proper subterm ordering. The ordering $\rhd_{R}$ is well-founded, but not necessarily a rewrite relation.[5] It contains both the rewrite relation $\rightarrow^{+}_{R}$ and the reduction ordering $\succ$.

Proofs are compared according to their complexity, using the lexicographic combination of the multiset extension of the ordering $\rhd_{R}$ (in the first component), the encompassment ordering $\rhd$ (in the second component), and the ordering $\rhd_{R}$ (in the last component). As usual, $\top$ and $\bot$ are assumed to be maximum and minimum elements with respect to each ordering. We denote this lexicographic ordering by $\succ^{p}$ and write $P \succsim_{\mathcal{P}} Q$ if $\gamma^{P} \succeq^{p} \gamma^{Q}$. In other words, we have $P \succ_{\mathcal{P}} Q$ if $\gamma^{P} \succ^{p} \gamma^{Q}$. The strict ordering $\succ_{\mathcal{P}}$ is well-founded.

The inference rules of $\mathcal{P}$ never increase the complexity of proofs.

Lemma 5.12. *Suppose* $L\,;C \vdash_{\mathcal{P}} L'\,;C'$ *and let* P *be a proof of an equation* $s' \approx t'$ *in* $R \cup C$. *Then there is a proof* P' *in* $R \cup C'$, *such that* $P \succsim_{\mathcal{P}} P'$, *if* $s' \approx t'$ *is an inductive theorem of* R, *and* $P \succeq_{\mathcal{P}} P'$, *if* $s' \approx t'$ *is not an inductive theorem.*

Proof. Suppose $L\,;C \vdash_{\mathcal{P}} L'\,;C'$ and let P be a proof in $R \cup C$. If P is a proof of an inductive theorem of R, then $\gamma^{P} \succeq^{p} (\top, \bot, \bot)$. Let P' be the empty proof. Clearly, we have $P \succsim_{\mathcal{P}} P'$.

On the other hand, suppose P is a ground proof $s' \leftarrow^{*}_{R} s\sigma \leftrightarrow^{\lambda}_{s\approx t} t\sigma \rightarrow^{*}_{R} t'$, where $s \simeq t$ is an equation in C and s' and t' are not equivalent in R. Since rewrite steps in R do not contribute to the complexity of a proof, we may assume, without loss of generality, that s' and t' are irreducible by R.

---

[5] The subterm ordering is not a rewrite relation. For example, x is a subterm of x^{-}, but $x + y$ is not a subterm of $x^{-} + y$.

If P is a proof in $R \cup C'$, let P' be P. If P is not a proof in $R \cup C'$, then C' is $(C \setminus \{s \approx t\}) \cup \{u \approx t\}$, where $s \succ u$ and either $s \leftrightarrow^+_{RUL} u$, or else $s \leftrightarrow^p_{v \approx w} u$ where $v \approx w$ is an equation in C, such that $s \rhd v$ and $v \succ w$. (The equation $s \approx t$ cannot be deleted, as it is not an inductive theorem.) Let u' be the normal form of $u\sigma$.

(a) If $u' \not\approx t'$, let P' be the proof $u' \leftarrow^*_R u\sigma \leftrightarrow^\lambda_{u \approx t} t\sigma \rightarrow^*_R t'$. Then the complexity of P' is the cost of the proof step $u\sigma \leftrightarrow^\lambda_{u \approx t} t\sigma$. We show that $P \succ_P P'$.

(i) If $s \succ t$, then $\gamma^P = (\{s\sigma\}, s, t\sigma)$. The first component of $\gamma^{P'}$ is $\{u\sigma, t\sigma\}$, $\{u\sigma\}$, or $\{t\sigma\}$, all of which are smaller than $\{s\sigma\}$ (with respect to the multiset extension of $\rhd_R$), so that $P \succ_P P'$.

(ii) If $t \succ s$, then $\gamma^P = (\{t\sigma\}, t, s\sigma) \succ_P (\{t\sigma\}, t, u\sigma) = \gamma^{P'}$.

(iii) If neither $s \succ t$ nor $t \succ s$, then $\gamma^P = (\{s\sigma, t\sigma\}, \perp, \perp)$, while the first component of $\gamma^{P'}$ is $\{u\sigma, t\sigma\}$ or $\{t\sigma\}$, both of which are smaller than $\{s\sigma, t\sigma\}$.

(b) If $u' = t'$, then $u' \neq s'$. Since $s \leftrightarrow^+_{RUL} u$ implies that s' and u' are identical, this case only arises if $s \leftrightarrow^p_{v \approx w} u$ for some equation $v \approx w$ in C, such that $s \rhd v$ and $v \succ w$. Let τ be a substitution, such that $s|_p = v\tau$ and $u|_p = w\tau$, and let v' and w' be the (ground) normal forms of $v\tau\sigma$ and $w\tau\sigma$, respectively. Note that $s' \neq u'$ implies $v' \neq w'$.

Let P' be the proof $v' \leftarrow^*_R v\tau\sigma \leftrightarrow^\lambda_{v \approx w} w\tau\sigma \rightarrow^*_R w'$. Since $v \succ w$, the complexity of P' is $(\{v\tau\sigma\}, v, w\tau\sigma)$. The complexity of P is $(\{s\sigma\}, s, t\sigma)$, if $s \succ t$; $(\{t\sigma\}, t, \sigma)$, if $t \succ s$; and $(\{s\sigma, t\sigma\}, \perp, \perp)$, otherwise. Considering the different cases and using the fact that $v\tau\sigma$ is a subterm of $s\sigma$ and $s \rhd v$, we conclude that $P \succ_P P'$.   Q.E.D.

Derivations for which proof normalization yields the desired result can be characterized by an appropriate fairness condition.

Definition 5.13. We say that C' is a *covering set* for C (or C' *covers* C) with respect to R and $\succ$ if, for every ground proof $P = s\sigma \leftrightarrow^\lambda_{s \approx t} t\sigma$ in C, such that (i) the substitution σ is irreducible by R, (ii) $s\sigma$ is reducible by R, and (iii) $t \not\succ s$, there exists a ground proof Q in C', such that $P \succ_P Q$.

Observe that if $s\sigma$ and $t\sigma$ are equivalent in R, then $P \succ_P Q$, for the empty proof Q. Consequently, the empty set is a covering set for any consistent set C. It is also evident from the definition that if C' covers C, then any superset of C' covers any subset of C. Moreover, if C'_1 covers C_1 and C'_2 covers C_2, then $C'_1 \cup C'_2$ covers $C_1 \cup C_2$.

Definition 5.14. A derivation $L_0 ; C_0 \vdash_P L_1 ; C_1 \vdash_P L_2 ; C_2 \vdash_P \cdots$ is said to be *fair* if the set $\bigcup_i C_i$ of all deduced equations covers the set C_∞ of all persisting conjectures.

A proof by consistency procedure is *fair* if it produces only fair derivations. With these definitions, we have:

Theorem 5.15. *If L_0 ; $C_0 \vdash_{\mathcal{P}} L_1$; $C_1 \vdash_{\mathcal{P}} L_2$; $C_2 \vdash_{\mathcal{P}} \cdots$ is a fair derivation, where C_0 is inconsistent with R, then some set C_i is provably inconsistent.*

Proof. Let L_0 ; $C_0 \vdash_{\mathcal{P}} L_1$; $C_1 \vdash_{\mathcal{P}} L_2$; $C_2 \vdash_{\mathcal{P}} \cdots$ be a fair derivation, where C_0 is inconsistent with R. We show that some set C_i is provably inconsistent. If C_0 is provably inconsistent, we are done. Let us assume C_0 is not provably inconsistent.

We first prove that whenever P is a proof $s' \leftarrow_R^+ s\sigma \leftrightarrow_{s \approx t}^\lambda t\sigma \rightarrow_R^* t'$ in $R \cup C_i$, such that s' and t' are distinct, irreducible ground terms, the substitution σ is irreducible, and $t \not\succ s$, then there is a proof P' in $R \cup C_j$, for some $j \geq 0$, such that $P \succ_{\mathcal{P}} P'$. By Lemma 5.12, this is true whenever $s \approx t \notin C_\infty$. Suppose the equation $s \approx t$ is in C_∞. By fairness $\bigcup_k C_k$ covers $\{s \approx t\}$. Thus there exists a ground instance $u\tau \approx v\tau$ of an equation $u \approx v$ in C_j, for some $j \geq 0$, such that $s\sigma \leftrightarrow_{s \approx t}^\lambda t\sigma \succ_{\mathcal{P}} u\tau \leftrightarrow_{u \approx v}^\lambda v\tau$. Then $P \succ_{\mathcal{P}} P'$, where P' is the proof $u' \leftarrow_R^* u\tau \leftrightarrow_{C_j} v\tau \rightarrow_R^* v'$ and u' and v' are normal forms of $u\tau$ and $v\tau$, respectively.

Let now P_{i_0} be a proof $s' \leftarrow_R^+ s\sigma \leftrightarrow_{C_0} t\sigma \rightarrow_R^* t'$, where s' and t' are distinct terms irreducible by R and σ is an irreducible substitution. (Since C_0 is not provably inconsistent, there exists such a proof.) Then there is a sequence of proofs P_{i_0}, P_{i_1}, P_{i_2}, ... where P_{i_j} is a proof in $R \cup C_{i_j}$ and $P_{i_{j-1}} \succ_{\mathcal{P}} P_{i_j}$, for all $j > 0$. Since $\succ_{\mathcal{P}}$ is well-founded this sequence must be finite. In other words, some proof P_{i_k} is of the form $s\sigma \leftrightarrow_{C_k} t\sigma \rightarrow_R^* t'$, where $s\sigma$ is irreducible and either $s \succ t$ or else $t\sigma$ is irreducible and distinct from $s\sigma$. But then C_{i_k} is provably inconsistent. Q.E.D.

Fair derivations always exist.

Lemma 5.16. *The set of all critical pairs of rules in R on equations in $C \cup C^{-1}$ is a covering set for C.*

Proof. Let C' be the set of all critical pairs of rules in R on equations in $C \cup C^{-1}$ and let P be a ground proof $s\sigma \leftrightarrow_{s \approx t}^\lambda t\sigma$ in C, where the substitution σ is irreducible by R, $s\sigma$ is reducible by R, and $t \not\succ s$. We have to show that there exists a ground proof Q in C', such that $P \succ_{\mathcal{P}} Q$. This trivially holds if $s\sigma$ and $t\sigma$ are equivalent in R. Let us assume this is not the case. Since σ is irreducible, but $s\sigma$ is reducible by R, there exist a critical pair $u \approx v$ of some rule in R on the equation $s \approx t$ and a substitution τ, such that $s\sigma \rightarrow_R u\tau$ and $t\sigma = v\tau$.

Let Q be the proof $u\tau \leftrightarrow_{u \approx v}^\lambda v\tau$. The complexity of Q is $(\{u\tau\}, u, v\tau)$, if $s \succ v$; $(\{v\tau\}, v, u\tau)$, if $v \succ u$; and $(\{u\tau, v\tau\}, \bot, \bot)$, otherwise. If $s \not\succ t$

(in addition to $t \not\succ s$), then the complexity of P is $(\{s\sigma, t\sigma\}, \bot, \bot)$. Since $s\sigma \succ_R u\tau$ and $t\sigma = v\tau$, we conclude that in this case $P \succ_{\mathcal{P}} Q$. On the other hand, if $s \succ t$, then the complexity of P is $(\{s\sigma\}, s, t\sigma)$. Since $s\sigma \succ_R u\tau$ and $s\sigma \succ_R t\sigma = v\tau$, we again have $P \succ_{\mathcal{P}} Q$. In sum, the set $CP(R, C \cup C^{-1})$ indeed covers C.   Q.E.D.

The lemma indicates that covering sets can be computed for all finite sets of equations C. As a consequence, we obtain:

Proposition 5.17. *Let C_0 be a finite set inconsistent with R. Then there is a derivation $L_0 ; C_0 \vdash_{\mathcal{P}} L_1 ; C_1 \vdash_{\mathcal{P}} L_2 ; C_2 \vdash_{\mathcal{P}} \cdots$ wherein some set C_i is provably inconsistent.*

In practice, fair derivation are constructed by employing some schema of marking those equations for which covering sets have already been computed. To guarantee fairness a covering set has to be computed for each unmarked equation. If the initial set of conjectures C_0 is inconsistent, then by the above theorem a provably inconsistent set will be produced eventually. If at any point in a derivation all equations are marked and no provably inconsistent set was encountered, then all deduced equations in $\bigcup_i C_i$ are inductive consequences of R. Of course, a procedure need not necessarily terminate for any consistent set, but may instead deduce ever more inductive consequences without obtaining a verifiably consistent set. The introduction of appropriate lemmas (e.g., by *generalization* as in Boyer and Moore 1979) in conjunction with simplification may be of help in obtaining a verifiably consistent set.

5.4.   Covering Sets

Computation of covering sets is usually based on critical pairs, but it is not always necessary to compute all critical pairs of R on $C \cup C'$. For instance, if $s \approx t$ is an equation with $s \succ t$, then $CP(R, \{s \approx t\})$ is a covering set for $\{s \approx t\}$. In other words, critical pairs which are obtained by superposing on the smaller term t are superfluous. This restriction conforms to the usual practice in completion procedures, but in proof by consistency procedures even more critical pairs can usually be eliminated.

Definition 5.18. (Fribourg 1989) A position p in a term t is said to be *inductively complete* (with respect to R) if $t|_p$ is not a variable and each ground term $t\sigma|_p$ is an instance of some left-hand side of R, whenever σ is irreducible.

In other words, if p is an inductively complete position in t, then each ground instance $t\sigma$ is reducible either within the variable part of t or at position p.

For example, take the rewrite system $R_0 = \{x + 0 \to x, x + S(y) \to S(x + y)\}$ and let s be the term $x + (y + z)$. The position $p = (2)$, referring to the subterm $y + z$, is inductively complete.

If p is an inductively complete position in a term s, then the set of all critical pairs obtained by superposing rules in R on the equation $s \approx t$ at position p in s is a covering set for $\{s \approx t\}$. As Fribourg (1989) points out, choosing an inductively complete position in a term, if one exists at all, essentially corresponds to selecting an "induction schema."

For example, the associativity of addition, $x + (y + z) \approx x + y) + z$, is an inductive consequence of the above set R. Let $\succ$ be the lexicographic path ordering based on a precedence $\succ_p$ in which $+ \succ_p S \succ_p 0$. We obtain a covering set for $\{x + (y + z) \approx (x + y) + z\}$ by superposing on the left-hand side $x + (y + z)$ at position $p = (2)$. There are two critical pairs, $x + y \approx (x + y) + 0$ and $x + (y + S(z)) \approx (x + y) + S(z)$, the first of which can be simplified to a trivial equation $x + y \approx x + y$, while the second can be simplified to a ground reducible equation $S(x + (y + z)) \approx S((x + y) + z)$, which is covered by $\{x + (y + z) \approx (x + y) + z\}$. Since the set of all deduced equations covers all persisting equations and no inconsistency has been derived, all deduced equations are inductive theorems of R.

In general, $\{s \approx t\}$ is a covering set for $\{u[s]_p \approx u[t]_p\}$, whenever $p \neq \lambda$. More precisely, whenever $u\sigma[s\sigma]_p \approx u\sigma[t\sigma]_p$ is an inconsistent ground instance, so is $s\sigma \approx t\sigma$, and in addition $u\sigma[s\sigma]_p \leftrightarrow^\lambda_{u[s]\approx u[t]} u\sigma[t\sigma]_p \succ_p s\sigma \leftrightarrow^\lambda_{s\approx t} t\sigma$.

In a similar vein, in theories with free constructors the set of equations $\{s_1 \approx t_1, \dots, s_n \approx t_n\}$ can be shown to be a covering set for any equation $f(s_1, \dots, s_n) \approx f(t_1, \dots, t_n)$, where f is a constructor symbol (cf. Huet and Hullot 1982). For if f is a free constructor, then a ground instance $f(s_1\sigma, \dots, s_n\sigma) \approx f(t_1\sigma, \dots, t_n\sigma)$ is inconsistent if and only if one of the equations $s_i\sigma \approx t_i\sigma$ is inconsistent. Moreover, in such theories any equation $f(s_1, \dots, s_m) \approx g(t_1, \dots, t_n)$, where f and g are different constructors, is inconsistent.

The notion of an inductively complete position can be generalized in an obvious way to sets of positions. A set $\{p_1, \dots, p_n\}$ of non-variable positions in a term t is said to be *inductively complete* if every ground instance $t\sigma$ is reducible by R at some position p_i, whenever σ is irreducible (cf. Küchlin 1989, Bündgen and Küchlin 1989).

Critical pair computations can often be restricted to inductively complete sets of positions even when an equation $s \approx t$ is unorientable. If s and

t are both ground reducible, then both $CP(R, \{s \approx t\})$ and $CP(R, \{t \approx s\})$ are covering sets for $\{s \approx t\}$. Thus, superposition may be restricted to either side of such an unorientable equation. (Proof by consistency differs from ordered completion in this respect.)

Example 5.19. Let R_0 be the rewrite system

$$
\begin{array}{ll|ll}
x + 0 & \rightarrow \ x & x * 0 & \rightarrow \ 0 \\
x + S(y) & \rightarrow \ S(x + y) & x * S(y) & \rightarrow \ x * y + x
\end{array}
$$

and $\succ$ be the recursive path ordering corresponding to a precedence $\succ_p$ in which $* \succ_p + \succ_p S \succ_p 0$. Furthermore let L_0 be the set of lemmas

$$
\begin{array}{ll}
x + y & \approx \ y + x \\
x + (y + z) & \approx \ (x + y) + z
\end{array}
$$

We prove that the set of conjectures

$$
\begin{array}{ll}
x * (y + z) & \approx \ x * y + x * z \\
x * y & \approx \ y * x \\
x * (y * z) & \approx \ (x * y) * z
\end{array}
$$

is consistent with R_0.

First observe that $x * (y + z) \succ x * y + x * z$. Hence we obtain a covering set for the distributivity equation by superposing on position $p = (2)$ of its left-hand side. There are two critical pairs, $x * y + x * 0 \approx x * y$ and $x * y + x * S(z) \approx x * S(y + z)$, the first of which can be simplified to a trivial equation $x * y \approx x * y$ and deleted. Since $x * y + x * S(z) \rightarrow_{R_0/L_0} (x * y + x * z) + x$ and $x * S(y + z) \rightarrow_{R_0} x * (y + z) + x \rightarrow_C (x * y + x * z) + x$, the second equation can also be simplified and deleted.

Superpositions on the commutativity axiom $x * y \approx y * x$ can be restricted to the left-hand side, as both $x * y$ and $y * x$ are ground reducible. This results in two critical pairs, $0 * x \approx 0$ and $S(y) * x \approx x * y + x$, both of which are ground reducible and can be used as rewrite rules. Computation of covering sets results in four new equations, three of which can be simplified to trivial equations. The remaining equation, $S(y) * x + S(y) \approx S(x) * y + S(x)$, can be simplified by $\rightarrow^+_{R_0/L_0}$ to $S(x * y + (x + y)) \approx S(y * x + (x + y))$, an equation which is already covered by the commutativity axiom.

Finally, to deal with the associativity axiom $x * (y * z) \approx (x * y) * z$, we superpose all rules in R_0 at the top-most position of the right-hand side of the equation and obtain a covering set of two equations, $x * y \approx x * (y * 0)$ and $(x * y) * z + (x * y) \approx x * (y * S(z))$. The first equation can be simplified to a trivial equation $x * y \approx x * y$; the second, to $(x * y) * z + (x * y) \approx$

$x * (y * z) + (x * y)$. The latter equation is covered by the associativity axiom.

In summary, all persisting equations are ground reducible and the set of all deduced equations covers the set of all persisting equations. Hence, all deduced equations, including the initial conjectures, are inductive theorems of R_0.

Example 5.20. Let R_1 be the rewrite system R_0 extended by the rules

$$\begin{array}{rcl} d(0) & \to & 0 \\ d(S(x)) & \to & S(S(d(x))) \end{array}$$

and let L_1 be L_0 plus all inductive theorems proved in the previous example. Then

$$d(x) \quad \to \quad x + x$$

is an inductive theorem of R_1.

Let us assume the lexicographic path ordering from the previous example is extended so that in the underlying precedence $d \succ_p *$. Then $d(x) \succ x + x$ and it suffices to superpose on $d(x)$ to get a covering set for the given conjecture. The critical pairs are $0 \approx 0 + 0$ and $S(S(d(x))) \approx S(x) + S(x)$, both of which can be reduced to trivial equations. For example,

$$S(S(d(x))) \to_{R_1} S(S(x + x)) \leftarrow_{R_1/L_1} S(x) + S(x).$$

Example 5.21. Let R_2 be the rewrite system R_0 extended by the rules

$$\begin{array}{rcl} x^0 & \to & S(0) \\ x^{S(y)} & \to & x * x^y \end{array}$$

and let $L_2 = L_1$. Again, we assume the lexicographic path ordering is extended by assigning highest precedence to the new function symbol. The conjecture

$$x^{y+z} \quad \approx \quad x^y * x^z$$

is covered by two critical pairs $x^y \approx x^y * x^0$ and $x^{S(y+z)} \approx x^y * x^{S(z)}$. We have

$$x^y \leftarrow^*_{R_2} x^y * 0 + x^y \leftarrow_{R_2} x^y * S(0) \leftarrow_{R_2} x^y * x^0$$

and

$$x^{S(y+z)} \to_{R_2} x * x^{y+z} \to_C x * (x^y * x^z) \leftarrow_{R_2/L_2} x^y * x^{S(z)}$$

which indicates that both critical pairs can be reduced to trivial equations. Since all persisting equations are covered, this establishes that $x^{y+z} \approx x^y * x^z$ is an inductive theorem of R_2.

Example 5.22. Let R_3 be the rewrite system R_0 extended by the rules

$$
\begin{array}{rcl}
sum(0) & \to & 0 \\
sum(S(x)) & \to & sum(x) + S(x)
\end{array}
$$

and let $L_3 = L_1$. The conjecture

$$
sum(x) + sum(x) \quad \approx \quad x * S(x)
$$

is covered by two critical pairs, $sum(0) + 0 \approx 0 * S(0)$ and $(sum(x) + S(x)) + sum(S(x)) \approx S(x) * S(S(x))$. Both critical pairs can be simplified to trivial equations. For example, we have

$$
\begin{array}{cl}
& (sum(x) + S(x)) + sum(S(x)) \\
\to_{R_3/L_3} & (S(x) + S(x)) + (sum(x) + sum(x)) \\
\to_C & (S(x) + S(x)) + (x * S(x)) \\
\leftarrow_{R_3/L_3} & (S(x) * S(x)) + S(x) \\
\leftarrow_{R_3/L_3} & S(x) * S(S(x))
\end{array}
$$

for the second critical pair. Again the conjecture is an inductive theorem.

Summary

We have presented an inductive theorem proving method that is based on the concept of proof by consistency. The essential components of this method are the computation of covering sets (e.g., via superposition on inductively complete positions) and a ground reducibility test to check for inconsistency. We have shown that proof by consistency is complete for disproving inductive theorems when applied to equational theories that can be represented as ground convergent rewrite systems. The method employs a linear deduction strategy and has the advantage over full inductive completion in that it does not fail when it encounters unorientable equations.

The ground convergence of the given theory R, it should be pointed out, is only needed to prove the refutation completeness of proof by consistency. The same techniques can also be applied to non-convergent sets of equations. Gramlich (1989, 1990a) formulates a more general proof transformation system for proof by consistency along the lines of ordered completion. An implementation of a proof by consistency procedure, along with a number of examples, in particular of theorems arising in the verification of sorting algorithms, is described in Gramlich (1990b). The relation of completion-based approaches to (Noetherian) induction has been studied by Reddy (1990).

Bibliography

[1] ANANTHARAMAN, S., AND HSIANG, J. 1990. Automated proofs of the Moufang identities in alternative rings. *J. Automated Reasoning* 6:79–109.

[2] ANANTHARAMAN, S., HSIANG, J., AND MZALI, J. 1989. SbReve2: A term rewriting laboratory with (AC-)unfailing completion. In *Proc. 3rd Int. Conf. on Rewriting Techniques and Applications*, Lect. Notes in Comput. Sci. 355, pp. 533–537. Berlin, Springer-Verlag.

[3] AVENHAUS, J. 1986. On the descriptive power of term rewriting systems. *J. Symbolic Computation* 2:109–122.

[4] AVENHAUS, J., AND MADLENER, K. 1990. *Term rewriting and equational reasoning*. In *Formal Techniques in Artificial Intelligence: A Source Book*, ed. R.B. Banerji. Amsterdam, Elsevier. To appear.

[5] BACHMAIR, L. 1987. Proof methods for equational theories. Ph.D. diss., University of Illinois, Urbana-Champaign.

[6] BACHMAIR, L. 1988. Proof by consistency in equational theories. In *Proc. Third Annual Symp. Logic in Computer Science*, pp. 228–233. Edinburgh, Scotland.

[7] BACHMAIR, L., AND DERSHOWITZ, N. 1986. Commutation, transformation, and termination. In *Proc. 8th Int. Conf. on Automated Deduction*, Lect. Notes in Comput. Sci. 230, pp. 5–20. Berlin, Springer-Verlag.

[8] BACHMAIR, L., AND DERSHOWITZ, N. 1987. Inference rules for rewrite-based first-order theorem proving. In *Proc. Second Annual Symp. on Logic in Computer Science*, pp. 331–337. Ithaca, New York.

[9] BACHMAIR, L., AND DERSHOWITZ, N. 1988. Critical pair criteria for completion. *J. Symbolic Computation* 6:1–18.

117

[10] BACHMAIR, L., AND DERSHOWITZ, N. 1989. Completion for rewriting modulo a congruence. *Theor. Comput. Sci.* 67:173–201.

[11] BACHMAIR, L., DERSHOWITZ, N., AND HSIANG, J. 1986. Orderings for equational proofs. In *Proc. Symp. on Logic in Computer Science*, pp. 346–357. Boston, Massachusetts.

[12] BACHMAIR, L., DERSHOWITZ, N., AND PLAISTED, D. A. 1989. *Completion without failure.* In *Resolution of Equations in Algebraic Structures (Vol. 2: Rewriting Techniques)*, ed. H. Aït-Kaci and M. Nivat, pp. 1–30. Boston, Academic Press.

[13] BACHMAIR, L., AND GANZINGER, H. 1990. On restrictions of ordered paramodulation with simplification. In *Proc. 10th Int. Conf. on Automated Deduction*, Lect. Notes in Comput. Sci. 449, pp. 427–441. Berlin, Springer-Verlag.

[14] BACHMAIR, L., AND GANZINGER, H. 1990. Completion of first-order clauses with equality by strict superposition. In *Proc. Second Int. Workshop on Conditional and Typed Rewriting Systems*, Lect. Notes in Comput. Sci. Berlin, Springer-Verlag. To appear.

[15] BACHMAIR, L., AND PLAISTED, D. A. 1985. Termination orderings for associative-commutative rewriting systems. *J. Symbolic Computation* 1:329–349.

[16] BAIRD, T. B., PETERSON, G. E., AND WILKERSON, R. W. 1989. Complete sets of reductions modulo associativity, commutativity and identity. In *Proc. 3rd Int Conf. Rewriting Techniques and Applications*, Lect. Notes in Comput. Sci. 355, pp. 29–44. Berlin, Springer-Verlag.

[17] BEN CHERIFA, A., AND LESCANNE, P. 1987. Termination of rewriting systems by polynomial interpretations and its implementation. *Sci. Comput. Program.* 9:137–159.

[18] BOUDET, A., JOUANNAUD, J.-P., AND SCHMIDT-SCHAUSS, M. 1988. Unification in free extensions of Boolean rings and Abelian groups. In *Proc. Third Annual Symp. on Logic in Computer Science*, pp. 121–130. Edinburgh, Scotland.

[19] BOYER, R. S., AND MOORE, J. S. 1979. *A computational logic.* New York, Academic Press.

[20] BROWN, T.C., JR. 1975. A structured design-method for specialized proof procedures. Ph.D. diss., California Institute of Technology, Pasadena.

[21] BUCHBERGER, B. 1979. A criterion for detecting unnecessary reductions in the construction of Groebner bases. In *Proc. EUROSAM '79*, Lect. Notes in Comput. Sci. 72, pp. 3–21. Berlin, Springer-Verlag.

[22] BUCHBERGER, B. 1984. A critical-pair/completion algorithm for finitely generated ideals in rings. In *Logic and Machines: Decision Problems and Complexity*, Lect. Notes in Comput. Sci. 171, pp. 137–161. Berlin, Springer-Verlag.

[23] BUCHBERGER, B. 1987. History and basic features of the critical-pair/completion procedure. *J. Symbolic Computation* 3:3–38.

[24] BÜNDGEN, R., AND KÜCHLIN, W. 1989. Computing ground reducibility and inductively complete positions. In *Proc. 3rd Int Conf. Rewriting Techniques and Applications*, Lect. Notes in Comput. Sci., pp. 59–75. Berlin, Springer-Verlag.

[25] COMON, H. 1990. Solving inequations in term algebras. In *Proc. Fifth Annual Symp. on Logic in Computer Science*, pp. 62–69. Philadelphia, Pennsylvania.

[26] DERSHOWITZ, N. 1982. Orderings for term-rewriting systems. *Theor. Comput. Sci.* 17:279–301.

[27] DERSHOWITZ, N. 1982. Applications of the Knuth-Bendix completion procedure. In *Proc. of the Seminaire d'Informatique Theorique*, pp. 95–111. Paris, France.

[28] DERSHOWITZ, N. 1985. Computing with rewrite systems. *Inf. Control* 64:122–157.

[29] DERSHOWITZ, N. 1987. Termination of rewriting. *J. Symbolic Computation* 3:69–116.

[30] DERSHOWITZ, N. 1989. *Completion and its applications*. In *Resolution of Equations in Algebraic Structures (Vol. 2: Rewriting Techniques)*, ed. H. Aït-Kaci and M. Nivat, pp. 31–85. Boston, Academic Press.

[31] DERSHOWITZ, N. 1990. A maximal-literal unit strategy for Horn clauses. In *Proc. Second Int. Workshop on Conditional and Typed Rewriting Systems*, Lect. Notes in Comput. Sci. Berlin, Springer-Verlag. To appear.

[32] DERSHOWITZ, N., AND JOUANNAUD, J.-P. 1990. Rewrite systems. In *Handbook of Theoretical Computer Science (Vol. B: Formal Models and Semantics)*. Amsterdam, North-Holland. To appear.

[33] DERSHOWITZ, N., AND MANNA, Z. 1979. Proving termination with multiset orderings. *Commun. ACM* 22:465–476.

[34] DERSHOWITZ, N., AND MARCUS, L. 1985. Existence and construction of rewrite systems. Unpublished manuscript.

[35] DERSHOWITZ, N., MARCUS, L., AND TARLECKI, A. 1988. Existence, uniqueness, and construction of rewrite systems. *SIAM J. Comput.* 17:629–639.

[36] DEVIE, H. 1990. When ordered completion fails. In *Proc. Second Int. Workshop on Conditional and Typed Rewriting Systems*, Lect. Notes in Comput. Sci. Berlin, Springer-Verlag. To appear.

[37] FAGES, F. 1987. Associative-commutative unification. *J. Symbolic Computation* 3:257–275.

[38] FAGES, F., AND HUET, G. 1986. Complete sets of unifiers and matchers in equational theories. *Theor. Comput. Sci.* 43:189–200.

[39] FRIBOURG, L. 1985. A superposition oriented theorem prover. *Theor. Comput. Sci.* 35:129–164.

[40] FRIBOURG, L. 1989. A strong restriction of the inductive completion procedure. *J. Symbolic Computation* 8:253–276.

[41] GALLIER, J., AND SNYDER, W. 1989. Complete sets of transformations for general E-unification. *Theor. Comput. Sci.* 67:203–260.

[42] GALLIER, J., AND SNYDER, W. 1990. Designing unification procedures using transformations: A survey. *EATCS Bull.* 40:273–326.

[43] GALLIER, J., SNYDER, W., NARENDRAN, P., AND PLAISTED, D. A. 1988. Rigid E-unification is NP-complete. In *Proc. Third Annual Symp. Logic in Computer Science*, pp. 218–227. Edinburgh, Scotland.

[44] GALLIER, J., SNYDER, W., AND RAATZ, S. 1989. Rigid E-unification and its application to equational matings. In *Resolution of Equations in Algebraic Structures (Vol. 1: Algebraic techniques)*, pp. 151–216. Boston, Academic Press.

[45] GANZINGER, H. 1987. A completion procedure for conditional equations. In *Conditional Term Rewriting Systems*, Lect. Notes in Comput. Sci. 308, pp. 62–83. Berlin, Springer-Verlag. To appear in *J. Symbolic Computation*.

[46] GOGUEN, J. A. 1980. How to prove algebraic inductive hypotheses without induction. In *Proc. 5th Conf. Automated Deduction*, Lect. Notes in Comput. Sci. 87, pp. 356–373. Berlin, Springer-Verlag.

[47] GOGUEN, J. A., AND MESEGUER, J. 1986. EQLOG: *Equality, types, and generic modules for logic programming*. In *Logic Programming: Functions, Relations, and Equations*, ed. D. DeGroot and G. Lindstrom, pp. 295–363. Englewood Cliffs, Prentice-Hall.

[48] GRAMLICH, B. 1989. Inductive theorem proving using refined unfailing completion techniques. Tech. Rep. SR-89-14, Universität Kaiserslautern, Germany.

[49] GRAMLICH, B. 1990. Completion based inductive theorem proving: A case study in verifying sorting algorithms. Tech. Rep. SR-90-04, Universität Kaiserslautern, Germany.

[50] GRAMLICH, B. 1990. Completion based inductive theorem proving: An abstract framework and its applications. In *Proc. ECAI-90*, pp. 314–319. Stockholm, Sweden.

[51] HENSCHEN, L. J., AND WOS, L. T. 1974. Unit refutations and Horn clauses. *J. ACM* 21:590–605.

[52] HSIANG, J. 1985. Refutational theorem proving using term-rewriting systems. *Artif. Intell.* 25:255–300.

[53] HSIANG, J., AND RUSINOWITCH, M. 1986. A new method for establishing refutational completeness in theorem proving. In *Proc. 8th Int. Conf. on Automated Deduction*, Lect. Notes in Comput. Sci. 230, pp. 141–152. Berlin, Springer-Verlag.

[54] HSIANG, J., AND RUSINOWITCH, M. 1987. On word problems in equational theories. In *Proc. 14th ICALP*, Lect. Notes in Comput. Sci. 267, pp. 54–71. Berlin, Springer-Verlag.

[55] HUET, G. 1980. Confluent reductions: Abstract properties and applications to term rewriting systems. *J. ACM* 27:797–821.

[56] HUET, G. 1981. A complete proof of correctness of the Knuth and Bendix completion algorithm. *J. Comput. Syst. Sci.* 23:11–21.

[57] HUET, G., AND HULLOT, J.-M. 1982. Proofs by induction in equational theories with constructors. *J. Comput. Syst. Sci.* 25:239–266.

[58] HUET, G., AND OPPEN, D. C. 1980. Equations and rewrite rules: A survey. In *Formal Language Theory: Perspectives and Open Problems*, ed. R. Book, pp. 349–405. New York, Academic Press.

[59] HULLOT, J. 1980. A catalogue of canonical term rewriting systems. Tech. Rep. CSL-113, SRI International, Menlo Park, California.

[60] JOUANNAUD, J.-P. 1983. Confluent and coherent equational term rewriting systems: Application to proofs in abstract data types. In *Proc. 8th Coll. on Trees in Algebra and Programming*, Lect. Notes in Comput. Sci. 59, pp. 269–283. Berlin, Springer-Verlag.

[61] JOUANNAUD, J.-P., AND KIRCHNER, C. 1990. Solving equations in abstract algebras: A rule-based survey of unification. Tech. Rep., Université de Paris Sud, Orsay.

[62] JOUANNAUD, J.-P., AND KIRCHNER, H. 1986. Completion of a set of rules modulo a set of equations. *SIAM J. Comput.* 15:1155–1194.

[63] JOUANNAUD, J.-P., AND KOUNALIS, E. 1986. Automatic proofs by induction in equational theories without constructors. In *Proc. Symp. Logic in Computer Science*, pp. 358–366. Boston, Massachusetts.

[64] JOUANNAUD, J.-P., AND MARCHE, C. 1990. Completion modulo associativity, commutativity and identity (AC1). In *Proc. Int. Symp. DISCO '90*, Lect. Notes in Comput. Sci. 429, pp. 111–120. Berlin, Springer-Verlag.

[65] KAMIN, S., AND LEVY, J.-J. 1980. Two generalizations of the recursive path ordering. Univ. of Illinois at Urbana-Champaign. Unpublished manuscript.

[66] KAPUR, D., MUSSER, D. R., AND NARENDRAN, P. 1988. Only prime superpositions need be considered in the Knuth-Bendix completion procedure. *J. Symbolic Computation* 6:19–36.

[67] KAPUR, D., NARENDRAN, P., ROSENKRANTZ, D. J., AND ZHANG, H. 1987. Sufficient-completeness, quasi-reducibility and their complexity. Tech. Rep., Dept. of Computer Science, SUNY at Albany.

[68] KAPUR, D., NARENDRAN, P., AND ZHANG, H. 1986. Proof by induction using test sets. In *Proc. 8th Int. Conf. on Automated Deduction*, Lect. Notes in Comput. Sci. 230, pp. 99–117. Berlin, Springer-Verlag. To appear in *J. Symbolic Computation*.

[69] KAPUR, D., NARENDRAN, P., AND ZHANG, H. 1987. On sufficient-completeness and related properties of term rewriting systems. *Acta Inf.* 24:395–415.

[70] KAPUR, D., AND SIVAKUMAR, G. 1984. Architecture of and experiments with rrl, a rewrite rule laboratory. In *Proc. NSF Workshop on the Rewrite Rule Laboratory*, pp. 33–56. Rensellaerville, New York.

[71] KIRCHNER, C. 1986. Computing unification algorithms. In *Proc. Symp. Logic in Computer Science*, pp. 206–216. Boston, Massachusetts.

[72] KIRCHNER, C. 1988. *Methods and tools for equational unification.* In *Resolution of Equations in Algebraic Structures (Vol. 2: Rewriting Techniques)*, ed. H. Aït-Kaci and M. Nivat, pp. 171–210. Boston, Academic Press.

[73] KLOP, J. W. 1987. Term rewriting systems: A tutorial. *Bulletin of the EATCS* 32:143–183.

[74] KLOP, J. W. 1990. Term rewriting systems: From Church-Rosser to Knuth-Bendix and beyond. In *Proc. 17th Int. Colloquium on Automata, Languages and Programming*, Lect. Notes in Comput. Sci. 443, pp. 350–369. Berlin, Springer-Verlag.

[75] KLOP, J. W. 1990. *Term rewriting systems.* In *Handbook of Logic in Computer Science*, ed. S. Abramsky, D. M. Gabbay, and T. S. E. Maibaum. Oxford, Oxford University Press. To appear.

[76] KNUTH, D. E. 1973. *The art of computer programming (Vol. 1: Fundamental algorithms).* Reading, Mass., Addison-Wesley. 2nd edn.

[77] KNUTH, D. E., AND BENDIX, P. B. 1970. Simple word problems in universal algebras. In *Computational Problems in Abstract Algebra*, ed. J. Leech, pp. 263–297. Oxford, Pergamon Press.

[78] KOUNALIS, E., AND RUSINOWITCH, M. 1987. On word problems in Horn logic. In *Proc. First Int. Workshop on Conditional Term Rewriting*, Lect. Notes in Comput. Sci. 204, pp. 390–399. Berlin, Springer-Verlag.

[79] KÜCHLIN, W. 1985. A confluence criterion based on the generalised Newman lemma. In *Proc. Eurocal '85*, Lect. Notes in Comput. Sci. 204, pp. 390–399. Berlin, Springer-Verlag.

[80] KÜCHLIN, W. 1986. Equational completion by proof simplification. Tech. Rep. 86-02, Dept. of Mathematics, ETH Zürich, Switzerland.

[81] KÜCHLIN, W. 1986. A generalized Knuth-Bendix algorithm. Tech. Rep. 86-01, Dept. of Mathematics, ETH Zürich, Switzerland.

[82] KÜCHLIN, W. 1989. *Inductive completion by ground proof transformation.* In *Resolution of Equations in Algebraic Structures (Vol. 2: Rewriting Techniques)*, ed. H. Aït-Kaci and M. Nivat, pp. 211–244. Boston, Academic Press.

[83] LANKFORD, D. S. 1975. Canonical inference. Tech. Rep. ATP-32, Dept. of Mathematics and Computer Science, University of Texas, Austin.

[84] LANKFORD, D. S. 1979. On proving term rewriting systems are Noetherian. Tech. Rep. MTP-3, Mathematics Department, Louisiana Tech. Univ., Ruston.

[85] LANKFORD, D. S. 1981. A simple explanation of inductionless induction. Tech. Rep. MTP-14, Mathematics Department, Louisiana Tech. Univ., Ruston.

[86] LANKFORD, D. S., AND BALLANTYNE, A. M. 1977. Decision procedures for simple equational theories with commutative axioms: Complete sets of commutative reductions. Tech. Rep. ATP-35, Dept. of Mathematics and Computer Science, University of Texas, Austin.

[87] LANKFORD, D. S., AND BALLANTYNE, A. M. 1977. Decision procedures for simple equational theories with permutative axioms: Canonical sets of permutative reductions. Tech. Rep. ATP-37, Dept. of Mathematics and Computer Science, University of Texas, Austin.

[88] LANKFORD, D. S., AND BALLANTYNE, A. M. 1977. Decision procedures for simple equational theories with associative-commutative axioms: Complete sets of associative-commutative reductions. Tech. Rep. ATP-39, Dept. of Mathematics and Computer Science, University of Texas, Austin.

[89] LANKFORD, D. S., AND BALLANTYNE, A. M. 1979. The refutation completeness of blocked permutative narrowing and resolution. In *Proc. Fourth Workshop on Automated Deduction*. Austin, Texas.

[90] LE CHENADEC, P. 1986. *Canonical forms in finitely presented algebras.* New York, Wiley.

[91] LESCANNE, P. 1983. Computer experiments with the REVE term rewriting system generator. In *Proc. 10th ACM Symp. on Principles of Programming Languages*, pp. 99–108. Austin, Texas.

[92] LESCANNE, P. 1989. Completion procedures as transition rules + control. In *Proc. TAPSOFT '89 (vol. 1)*, Lect. Notes in Comput. Sci. 351, pp. 28–41. Berlin, Springer-Verlag.

[93] MARTELLI, A., AND MONTANARI, U. 1982. An efficient unification algorithm. *ACM Trans. Program. Lang. Syst.* 4:258–282.

[94] MARTIN, U., AND NIPKOW, T. 1990. Ordered rewriting and confluence. In *Proc. 10th Int. Conf. on Automated Deduction*, Lect. Notes in Comput. Sci. 449, pp. 366–380. Berlin, Springer-Verlag.

[95] METIVIER, Y. 1983. About the rewriting systems produced by the Knuth-Bendix completion algorithm. *Inf. Process. Lett.* 16:31–34.

[96] MUSSER, D. R. 1980. On proving inductive properties of abstract data types. In *Proc. 7th ACM Symp. on Principles of Programming Languages*, pp. 154–162. Las Vegas, Nevada.

[97] MZALI, J. 1986. Methodes de filtrage equationnel et de preuve automatique de theoremes. Ph.D. diss., Université de Nancy.

[98] O'DONNELL, M. J. 1985. *Equational logic as a programming language.* Cambridge, Massachusetts, MIT Press.

[99] OHSUGA, A., AND SAKAI, K. 1986. Metis: A term rewriting system generator. Tech. Rep., ICOT Research Center, Tokyo, Japan.

[100] PATERSON, M. S., AND WEGMAN, M. N. 1978. Linear unification. *J. Comput. Syst. Sci.* 16:158–167.

[101] PAUL, E. 1986. On solving the equality problem in theories defined by Horn clauses. *Theor. Comput. Sci.* 44:127–153.

[102] PEDERSEN, J. 1984. Confluence methods and the word problem in universal algebra. Ph.D. diss., Emory University.

[103] PETERSON, G. E. 1983. A technique for establishing completeness results in theorem proving with equality. *SIAM J. Comput.* 12:82–100.

[104] PETERSON, G. E. 1990. Complete sets of reductions with constraints. In *Proc. 10th Int. Conf. on Automated Deduction*, Lect. Notes in Comput. Sci. 449, pp. 381–395. Berlin, Springer-Verlag.

[105] PETERSON, G. E., AND STICKEL, M. E. 1981. Complete sets of reductions for some equational theories. *J. ACM* 28:233–264.

[106] PLAISTED, D. A. 1985. Semantic confluence tests and completion methods. *Inf. Control* 65:182–215.

[107] PLOTKIN, G. 1972. *Building-in equational theories.* In *Machine Intelligence 7*, ed. B. Meltzer and D. Michie, pp. 73–90. New York, Wiley.

[108] REDDY, U. 1989. Rewriting techniques for program synthesis. In *Proc. 3rd Int Conf. on Rewriting Techniques and Applications*, Lect. Notes in Comput. Sci. 355, pp. 388–403. Berlin, Springer-Verlag.

[109] REDDY, U. 1990. Term rewriting induction. In *Proc. 10th Int. Conf. on Automated Deduction*, Lect. Notes in Comput. Sci. 449, pp. 162–177. Berlin, Springer-Verlag.

[110] ROBINSON, G. A., AND WOS, L. T. 1969. Paramodulation and theorem proving in first order theories with equality. In *Machine Intelligence 4*, ed. B. Meltzer and D. Michie, pp. 133–150. New York, American Elsevier.

[111] ROBINSON, J. A. 1965. A machine-oriented logic based on the resolution principle. *J. ACM* 12:23–41.

[112] RUSINOWITCH, M. 1991. Theorem proving with resolution and superposition: An extension of the Knuth and Bendix procedure as a complete set of inference rules. *J. Symbolic Computation.* To appear.

[113] SIEKMAN, J. 1984. Universal unification. In *Proc. 7th Int. Conf. on Automated Deduction*, Lect. Notes in Comput. Sci. 170, pp. 1–42. Berlin, Springer-Verlag.

[114] SLAGLE, J. R. 1974. Automated theorem proving for theories with simplifiers, commutativity, and associativity. *J. ACM* 21:622–642.

[115] STICKEL, M. E. 1981. A unification algorithm for associative-commutative functions. *J. ACM* 28:423–434.

[116] WINKLER, F. 1984. The Church-Rosser property in computer algebra and special theorem proving: An investigation of critical-pair/completion algorithms. Ph.D. diss., Johannes Kepler University, Linz, Austria.

[117] WINKLER, F. 1985. Reducing the complexity of the Knuth-Bendix completion algorithm: A 'unification' of different approaches. In *Proc. Eurocal '85*, Lect. Notes in Comput. Sci. 204, pp. 378–389. Berlin, Springer-Verlag.

[118] WINKLER, F., AND BUCHBERGER, B. 1983. A criterion for eliminating unnecesssary reductions in the Knuth-Bendix algorithm. In *Proc. Coll. on Algebra, Combinatorics and Logic in Computer Science*. Györ, Hungary.

[119] ZHANG, H., AND KAPUR, D. 1989. Consider only general superpositions in completion procedures. In *Proc. 3rd Int Conf. on Rewriting Techniques and Applications*, Lect. Notes in Comput. Sci. 355, pp. 513–527. Berlin, Springer-Verlag.

Index

Progress in Theoretical Computer Science

Editor
Ronald V. Book
Department of Mathematics
University of California
Santa Barbara, CA 93106

Progress in Theoretical Computer Science is a series that focuses on the theoretical aspects of computer science and on the logical and mathematical foundations of computer science, as well as the applications of computer theory. It addresses itself to research workers and graduate students in computer and information science departments and research laboratories, as well as to departments of mathematics and electrical engineering where an interest in computer theory is found.

The series publishes research monographs, graduate texts, and polished lectures from seminars and lecture series. We encourage preparation of manuscripts in some form of TeX for delivery in camera-ready copy, which leads to rapid publication, or in electronic form for interfacing with laser printers or typesetters.

Proposals should be sent directly to the Editor, any member of the Editorial Board, or to: Birkhäuser Boston, 675 Massachusetts Avenue, Cambridge, MA 02139.

Progress in Theoretical Computer Science